Springer Series in **Materials Science** 9

Edited by U. Gonser

Springer Series in **Materials Science**

Editors: U. Gonser · A. Mooradian · K. A. Müller · M. B. Panish · H. Sakaki
Managing Editor: H. K. V. Lotsch

Volume 1 **Chemical Processing with Lasers**
By D. Bäuerle

Volume 2 **Laser-Beam Interactions with Materials**
Physical Principles and Applications
By M. von Allmen

Volume 3 **Laser Processing of Thin Films and Microstructures**
Oxidation, Deposition and Etching of Insulators
By I. W. Boyd

Volume 4 **Microclusters**
Editors: S. Sugano, Y. Nishina, and S. Ohnishi

Volume 5 **Graphite Fibers and Filaments**
By M. S. Dresselhaus, G. Dresselhaus,
K. Sugihara, I. L. Spain, and H. A. Goldberg

Volume 6 **Elemental and Molecular Clusters**
Editors: G. Benedek, T. P. Martin, and G. Pacchioni

Volume 7 **Molecular Beam Epitaxy** Fundamentals and Current Status
By M. A. Herman and H. Sitter

Volume 8 **Physical Chemistry of, in and on Silicon**
By G. F. Cerofolini and L. Meda

Volume 9 **Tritium and Helium-3 in Metals**
By R. Lässer

Volume 10 **Computer Simulation of Ion – Solid Interactions**
By W. Eckstein

Volume 11 **Mechanisms of High Temperature Superconductivity**
Editors: H. Kamimura and A. Oshiyama

Volume 12 **Laser Technology in Microelectronics**
Editors: S. Metev and V. P. Veiko

Volume 13 **Semiconductor Silicon**
Materials Science and Technology
Editors: G. Harbeke and M. J. Schulz

R. Lässer

Tritium and Helium-3 in Metals

With 77 Figures

Springer-Verlag Berlin Heidelberg New York
London Paris Tokyo Hong Kong

Dr. Rainer Lässer

Institut für Festkörperforschung, Kernforschungsanlage Jülich,
D-5170 Jülich, FRG

Present address: JET Joint Undertaking, Abingdon,
Oxfordshire, OX14 3EA, United Kingdom

ISBN-13: 978-3-642-73512-7 e-ISBN-13: 978-3-642-73510-3
DOI: 10.1007/978-3-642-73510-3

Library of Congress Cataloging-in-Publication Data. Lässer, Rainer, 1950– Tritium and helium-3 in metals / Rainer Lässer. p. cm. – (Springer series in materials science ; v. 9) Includes bibliographical references. ISBN-13: 978-3-642-73512-7 1. Metals – Tritium content. 2. Metals – Helium content. I. Title. II. Series. TN690.L1778 1989 620.1'6 – dc20 89-19720

Softcover reprint of the hardcover 1st edition 1989

2154/3150-543210 – Printed on acid-free paper

Preface

Hydrogen can behave as an alkaline metal or a halogen and can react with nearly all elements of the periodic table. This explains the large number of metal hydrides.

Since T. Graham's first observation of the absorption of hydrogen in palladium in 1866 the behaviour of hydrogen in metals has been studied very extensively. The interest was motivated by the possible application of metal-hydrogen systems in new technologies (e.g., moderator material in nuclear fission reactors, reversible storage material for thermal energy and large amounts of hydrogen) and by the fact that metal hydrides show very exciting physical properties (e.g., superconductivity, quantum diffusion, order-disorder transitions, phase diagrams, etc.). Many of these properties have been determined for the stable hydrogen isotopes H and D in various metals. In comparison, very little is known about the behaviour of the radioactive isotope tritium in metals.

This book is a first attempt to summarize part of the knowledge of tritium gained in the last few years. In addition to the task of presenting the properties of tritium in metals, I have tried to compare these data with those of protium and deuterium. Furthermore, helium-3 is connected inseparably with tritium via the tritium decay. Therefore one chapter of this book is solely devoted to the curious properties of helium in metals caused mainly by its negligible solubility.

The book also treats general aspects of tritium such as its inventory on earth, its natural and man-made production paths, its release to the environment, incorporation in the body, handling, etc.

It was not my intention to present all the data on tritium in metals and I am aware of the fact that I have not discussed such interesting problems as the application of hydrogen-isotope separation in metals, the use of T in nuclear warheads to boost the energy release, etc. I would like to apologize to all colleagues whose work I did not mention.

I would like to thank D. Beyer, A. Dombra, R. Feenstra, T.B. Flanagan, R. Griessen, J.L. Hemmerich, W. Jäger, P. Jung, G. Kögel, R. Messer, W.A. Oates, G.L. Powell, H. Schober, T. Schober, G. Thomas, H. Trinkaus, J.-M. Welter and H. Wenzl for helpful discussions. The technical assistance of K. Bickmann, H.J. Bierfeld, K.-H. Klatt and D. Triefenbach was appreciated. The expert typing of the manuscript and the drawings were done by L. Jansen and K.-H. Stange. Special thanks are due to T. Schober, and K. Walter for reading the manuscript, to H. Wenzl for his continuous support during the years 1981–1988 and to H. Lotsch for his help in publishing this book.

Finally, I would like to thank my wife Elli for her understanding and patience because the main part of this book was prepared in my (our) own time.

April 1989
Harwell OX11 OPT
U.K.

R. Lässer

Contents

1. Introduction

This introductory chapter begins with a brief description of the discovery of tritium, before outlining some important properties of all of the hydrogen isotopes. Some essential background information about hydrogen and tritium in metals is included, too.

1.1 Brief Historical Survey

The superheavy and radioactive hydrogen isotope tritium was discovered in 1939 by *Alvarez* and *Cornog* [1.1]. Seven years later *Libby* [1.2] found the production of tritium by natural processes. In addition to these two simple statements, a few further historical details about the production and discovery of tritium are noteworthy. Very important prerequisites for the artificial fabrication of tritium were the development of particle accelerators and of the method for producing heavy water in about 1932. In 1934 tritium was generated for the first time in the Cavendish Laboratory in a group headed by *Lord Rutherford*, see [1.3]. This group used the accelerator at Cambridge to bombard a heavy-water target with accelerated deuterons from heavy hydrogen. They observed two reaction products, tritium and helium-3, stemming from the reactions $D+D \rightarrow T+H + 3.98$ MeV and $D+D \rightarrow {}^3He+n + 3.5$ MeV. At that time it was not known which of the two nuclides was the radioactive one. *Lord Rutherford* made the wrong guess, thinking that tritium is the stable isotope. He tried to enrich tritium by further electrolysis of heavy water and started a collaboration with the only deuterium producer at that time, a Norwegian heavy-water plant. The final product was analysed by a very sensitive mass spectrometer, and no mass 3 was found.

Alvarez, having heard of *Lord Rutherford's* work, tried to do the reverse and discovered helium-3 by analysing natural helium gas with a mass spectrometer. This clearly demonstrated that helium-3 has to be a stable isotope or, at least, that it has a very long half-life. As one of the next steps, *Alvarez* and *Cornog* showed that tritium is radioactive by bombarding lithium with neutrons and analysing the gaseous hydrogen in an ion chamber. After the second World War, *Libby* [1.2] decided to look again for natural tritium. He followed the steps of *Lord Rutherford* and succeeded in obtaining the same samples used by *Lord Rutherford* about 13 years earlier. *Libby* found such a high tritium activitiy in these samples, using Geiger counters, that he had to dilute them by a factor of a thousand. Later, *Libby* asked the directors of the Norwegian heavy-water plant to produce a concentrate by electrolyzing heavy water, because there was a possibility that the tritium observed in *Lord Rutherford's* sample might have been generated by neutron activation of heavy water. The analyses from the new concentrate gave the same answer, and the existence of natural tritium was established. It seems strange that *Lord Rutherford* failed to discover tritium, although his laboratory was certainly equipped with instruments capable of measuring the tritium activity in the sample. This is just one example of how things can go in life and physics. On the other hand, in addition to his many other important contributions to physics, he will be known as the man who generated tritium for the first time. After this brief historical flashback [1.4, 5], let us consider the behaviour of tritium and its decay product, helium-3, in metals.

1.2 Hydrogen Isotopes

Considering readily accessible particles carrying positive elementary charge as members of the family of hydrogen isotopes, the following generalized hydrogen isotopes are known: positronium: Ps (positron: e^+); muonium: Mu (muon: μ^+); protium: ^1H (proton: p); deuterium: D = ^2H (deuteron: d); tritium: T = ^3H (triton: t); ^4H and ^5H; where the name in parentheses stands for the positively charged particle. With the exception of ^4H and ^5H [1.6] these isotopes play important roles in physics. The study of solids with positrons is now a well-established technique for obtaining information about defects and electronic properties [1.7]. From the point of view of its mass, muonium is a very interesting hydrogen isotope. Its mass is one ninth of the protium mass. The vibrational energy in metals is expected to be 3 times higher than that of ^1H. Owing to the short lifetime mainly kinetic measurements of the mobility of muonium and determinations of the occupied sites in metals have been possible [1.8-10]. The natural hydrogen isotopes H, D and T are commercially available in large quantities and high purity, and are of importance as materials for research and technology. Various scenarios for the future predict that hydrogen isotopes will play an important role as energy carriers and as a source of nuclear energy in fusion reactors.

In the remainder of this introductory chapter, properties of protium, deuterium and tritium in metals will be presented. A few data on the nucleus of these isotopes are listed in Table 1.1 [1.11-13]. The other hydrogen "isotopes" mentioned above will not be discussed here.

Special properties of tritium are:

- Tritons are the nuclei with the largest nuclear gyromagnetic ratio [1.13]. Therefore, the triton is the most sensitive nucleus for nuclear

Table 1.1. Properties of the nuclei of the natural hydrogen isotopes [1.11-13]

Atom	Protium $^{1}H=p+e^-$	Deuterium $D = {}^{2}H=d+e^-$	Tritium $T = {}^{3}H=t+e^-$
nucleus	Proton p	Deuteron d	Triton t
Rest mass m [kg]	$1.6726\cdot10^{-27}$	$3.3436\cdot10^{-27}$	$5.0061\cdot10^{-27}$
Spin	1/2	1	1/2
Magnetic moment $[J\cdot T^{-1}]$	$1.41062\cdot10^{-26}$	$4.33066\cdot10^{-27}$	$1.50457\cdot10^{-26}$
Gyromagnetic ratio $[rad\cdot s^{-1}\cdot T^{-1}]$	$2.6752\cdot10^{8}$	$0.4106\cdot10^{8}$	$2.853\cdot10^{8}$
Quadrupole moment $[em^2]$	0	$2.77\cdot10^{-31}$	0
Halflife $T_{1/2}$ [years]	stable	stable	12.323
Incoherent neutron cross-section, σ_{inc} [barns]	80.20 ± 0.06	2.04 ± 0.3	$0.0\,^{+0.1}_{-0.8}$
Bound scattering cross-section $\sigma_{inc} + \sigma_{coh}$ [barns]	81.96 ± 0.06	7.64 ± 0.03	2.3 ± 0.7
NMR resonance frequency at 1.114 T [MHz]	90.0	13.8	96.0
Relative NMR sensitivity at constant field	1.0	$9.65\cdot10^{-3}$	1.21
Natural abundance [%]	99.984	0.0156	10^{-16}

magnetic resonance (NMR). Its sensitivity is even 1.21 times [1.13] larger than the one of the proton. Tritium NMR spectroscopy is one of the best methods to analyse tritium-labelled compounds. The method is nondestructive, rapid and can also be used to study the behaviour of tritium and its decay product helium-3 in metals.

- Tritium extends the number of the usually employed stable hydrogen isotopes and offers a further possibility to test theories.
- The H, D and T atoms show the largest relative mass changes of all isotopic species (an exception is muonium). Therefore, mass-dependent physical properties should show the largest changes for these isotopes.
- The detection limit for tritium is far lower than for H or D because, owing to the radioactive decay of T, nuclear-physical methods can be used. Thus tritium is employed as label and tracer for hydrogen, and is the most versatile beta emitting radionuclide for use in chemical and biological research.

Table 1.2. Characteristic data of tritium. The data are partly taken from [1.18]

atomic weight:	3.01605g
radiation:	β^- decay (100%)
half life: $T_{1/2}$:	12.323y (1y=365.25d)
decay constant: $\lambda=\ln 2/T_{1/2}$:	$1.782 \cdot 10^{-9} s^{-1}$
decay product:	^3He (atomic weight: 3.01603g)
maximum β energy:	18.582keV
mean β energy:	5.685keV
heat produced by tritium:	0.324W/g $\hat{=}$ 1.954W/molT_2
ionization energy: $T \rightarrow T^+ + e^-$:	13.595eV
dissociation energy: $T_2 \rightarrow 2T$:	4.591eV
tritium activity per gram:	355.7TBq (9615Ci)
tritium activity per mol T_2:	2146TBq (58.00kCi)
tritium activity in 1cm^3 at 1bar and 293K:	88.1GBq (2.38Ci)
tritium activity of 1g T_2O:	97.42TBq (2633Ci)
tritium activity of 1g DTO:	51.0TBq (1379Ci)
tritium activity of 1g HTO:	53.6TBQ (1449Ci)

- T will also play a very important role in future fusion reactors.

A few characteristic data for tritium are listed in Table 1.2.

1.3 Hydrogen in Metals

Due to the unfilled electron shell and the intermediate electronegativity compared with other elements, the hydrogen atom reacts with almost all elements of the periodic table. The reaction products of the hydrogen isotopes with metals (M) are called metal protides (MH_r), metal deuterides (MD_r) and metal tritides (MT_r). Their binding character can be ionic, covalent or metallic. The metal hydrides are of great interest given their interesting properties which include:

- extremely high mobility of the hydrogen atom,
- phase transitions between ordered and disordered structures in a wide temperature and concentration range,
- increase or decrease of the superconducting transition temperature upon alloying with H, D and T,
- large hydrogen storage capacity,
- hydrogen embrittlement, etc.

Many of these properties have been reviewed in books [1.14] and discussed in metal-hydrogen conferences [1.15] for H and D. Not one chapter of these reviews has been devoted to the behaviour of tritium in metals until now. This book tries to close the gap of information by summarizing recently determined properties of tritium in metals. Fusion relevant tritium technology has also been presented in a few conferences [1.16].

1.4 Helium in Metals

In contrast to the hydrogen isotopes, He with its closed electron shell is a very stable atom and forms only a limited number of compounds with mainly van der Waals binding character. It shows an extremely low solubility in metals. In most studies of the He behaviour in metals, samples doped with He by α implantation have been used. Another technique (the so-called "tritium trick"), uses tritium decay: the tritium atom decays with a half-life of $T_{1/2}$ = 12.323 years (1 year ≃ 365.25 days) [1.17] according to

$$T \rightarrow {}^3He^+ + \beta^- + \bar{\nu}_e + 18.582\ \text{keV} \tag{1.1}$$

into a positively charged He ion, an electron β^-, and an antineutrino $\bar{\nu}_e$. Given the transmutation of tritium into ^{3}He, metal tritides have to be considered as time-dependent ternary alloys $MT_{r_0-c}He_c$ with a helium concentration c that increases as

$$c(t) = r_0\,[1 - \exp(-t\ \ln 2/T_{1/2})]\ . \tag{1.2}$$

The decreasing tritium concentration is expressed by (r_0-c), where r_0 is the initial tritium concentration. The concentrations r_0 (c) are given as the ratio of T (He) atoms to metal atoms. Changes of physical properties due to ^{3}He can be studied in the ternary alloy itself, or after removal of the tritium. The tritium can be removed by degassing the ternary alloy at high temperatures in a vacuum system, or by spot welding it to other metals with a larger affinity for tritium.

1.5 Organization of the Book

The present book, the purpose of which is to review the behaviour of tritium and helium-3 in metal tritides, is organized as follows: In Chap.2 a few general aspects of tritium such as its production, natural abundance, associated risks, and handling in future fusion reactors are discussed. Chapter 3 describes possible methods of charging metals with tritium. A variety of techniques for analysing tritium in metals are described in Chap.4. The isotope dependence of such differing physical properties as solubility, phase boundaries, superconductivity, diffusion, optical phonons, etc. are discussed in Chap.5. The behaviour of ^{3}He in metals is treated in Chap.6.

The properties of the hydrogen molecules H_2, HD, HT, D_2, DT and T_2 and of the tritium decay product ^{3}He are mentioned only briefly and where necessary; they have been reviewed very recently in an excellent book by Souers [1.18]. The isotopic dependence of a few physical quantities is shown in Table 1.3 for the hydrogen molecules.

Table 1.3. Characteristic properties of the 6 hydrogen molecules [1.18, 19]

	H_2	HD	HT	D_2	DT	T_2
Molecular weight [g]	2.015650	3.021927	4.023875	4.028204	5.030152	6.032100
Triple point temp. [K]	13.804	16.60	17.70	18.69	19.79	20.62
Triple point pressure [bar]	0.0703	0.1237	0.1458	0.1713	0.2008	0.216
Critical point temp.[K]	32.976	35.91	37.13	38.262	39.42	40.44
Critical point Pressure [bar]	12.93	14.84	15.71	16.5	17.73	18.50
Ionization energy [eV]	15.43	15.44	15.45	15.47	15.47	15.49
Dissociation energy [eV]	4.478	4.514	4.527	4.556	4.573	4.591
First excited vibrational level [K] (Einstein temp.)	5986	5225	4940	4307	3948	3548

Only information available in the open literature is reviewed. No attempt is made to present all the information published over the years and the selection given here is inevitably highly subjective.

2. General Aspects of Tritium

In this chapter, important background information about tritium, such as its production, toxicity, inventories on earth, the discharge to the environment and its handling, are discussed.

2.1 Production of Tritium

The main causes of tritium formation are cosmic rays, ternary fission of uranium and plutonium isotopes, and neutron activation.

2.1.1 Natural Tritium Production

The natural production involves processes such as

$$^{14}N + n \rightarrow {}^{12}C + T \quad \text{and} \tag{2.1a}$$

$$^{16}O + n \rightarrow {}^{14}N + T \tag{2.1b}$$

in the atmosphere. Furthermore, T is also generated by high-energy protons. Most of the natural tritium is produced by primary cosmic-ray reactions in the atmosphere. The production rates vary with altitude, latitude and the eleven-year solar cycles. The world inventory of natural tritium and the corresponding production rate are about $1.3 \cdot 10^6$ TBq (35 MCi) and $7.4 \cdot 10^4$ TBq/annum (2 MCi/a) [2.1], respectively. (T is an abbreviation for Tera and means 10^{12}.) The Becquerel (Bq) is the new unit for activity introduced in 1986: 3.7 TBq $\hat{=}$ 100 Ci; 3.7 TBq tritium corresponds to 10.399 mg of tritium. A further unit is the so-called Tritium Unit (TU) which corresponds to 1 tritium atom per 10^{18} ^1H atoms or to 0.120 Bq/litre water or to 1.08 Bq/kg hydrogen. The natural tritium inventory on earth is about 3.7 kg. Most of which exists in the oxidized form. Concentrations of natural tritium in ocean water and in rain are about 0.11 Bq (3pCi) and 0.15 Bq (4pCi) per litre, respectively [2.2].

Table 2.1. Ternary fission yields of tritium for fission of the listed nuclides [2.3]

Nuclides	Range of fission yield [T atoms/fission]
Uranium-235	$(0.8 - 1.32)\cdot 10^{-4}$
Uranium-238	$(0.68 - 0.91)\cdot 10^{-4}$
Plutonium-239	$(1.34 - 1.8)\cdot 10^{-4}$

2.1.2 Tritium Production in Nuclear Reactors

For ternary fission, the most important nuclides for tritium production in conventional nuclear reactors are shown with the corresponding fission yield in Table 2.1 [2.3]. For neutron activation, the nuclides and cross-sections are listed in Table 2.2 [2.3]. A part of the tritium produced in the nuclear fuel (Table 2.1) can leave the fuel element. In the case of stainless-steel cladding, tritium can permeate easily. Only about 0.1% of the tritium produced diffuses through Zircaloy cladding and reaches the coolant. The total tritium production by ternary fission in all nuclear reactors was estimated to be $10.5\cdot 10^4$ TBq (2.84 MCi) [2.3] during the year 1984.

The reactions listed in Table 2.2 are important in control rods, chemical additives to the coolant and the coolant itself. In heavy-water reactors, the tritium production in the primary coolant stems mainly from neutron activation of deuterium atoms, whereas in pressurized water reactors it mainly results from the activation of dissolved boron and lithium hydroxide. Lithium present as an impurity in the graphite moderator is an important source of tritium in gas-cooled reactors. The global production of T by n activation was estimated to be $5.5\cdot 10^3$ TBq (0.15 MCi) [2.3] during the year 1984. Almost no data are available for tritium production plants. One such plant is Savannah River in USA where T is produced by n activation of Al-Li alloys.

2.1.3 Tritium Production in Future Fusion Reactors

Due to the limited amount of about 3.7 kg of natural tritium (Sect. 2.1.1) the neutron activation processes of lithium listed in Table 2.2 will be employed in future fusion reactors because the most promising fusion reaction

$$D + T \rightarrow {}^4He\ (3.52\,MeV) + n\ (14.1\,MeV) \tag{2.2}$$

needs T as fuel. Because some of the 14.1 MeV neutrons will be absorbed in the walls of fusion reactors, neutron (n) multipliers such as Pb and Be with (n,2n) processes have to be used. A disadvantage of this neutron multiplication is that the two neutrons may have insufficient energy to activate

Table 2.2. Neutron activation reactions for tritium formation [2.3]

reactions	$\sigma\,[10^{-28}\ cm^2]$
$^2D + n \rightarrow T + \gamma$	$3.16 \cdot 10^{-4}$
$^6Li + n \rightarrow T + {}^4\alpha$	693
$^7Li + n \rightarrow T + {}^4\alpha + n$	$5.16 \cdot 10^{-2}$
$^{10}B + n \rightarrow {}^7Li + {}^4\alpha$	3060
$^{10}B + n \rightarrow T + 2\,{}^4\alpha$	1.27

^{7}Li because neutrons with energies above 2.87 MeV are required. On the other hand, neutrons of any energy can activate ^{6}Li, but this isotope has an abundance of only 7.5% in natural Li, and also has a low neutron capture cross-section (Table 2.2).

The tritium required for start-up of a fusion reactor has to be supplied by nuclear reactors or by tritium production plants. More information about T in fusion reactors will be given in Sect.2.4.

2.1.4 Tritium Production by Nuclear Explosions

Tritium is produced in fission and in fusion explosions. In fission detonations, about 25 TBq (675Ci) tritium/Mton TNT equivalent [2.4], are generated assuming a fission yield of $1 \cdot 10^{-4}$ (Table 2.1) and $1.4 \cdot 10^{26}$ fissions/Mton TNT equivalent [2.5]. The tritium yield by n activation is estimated to be about $7 \cdot 10^3$ TBq (0.2MCi)/Mton equivalent [2.6].

In fusion bombs, a fraction of the tritium produced by n activation of ^{6}Li does not react with deuterium according to (2.2), but escapes instead into the atmosphere. The estimated values range from ($2.6 \cdot 10^5$ to $1.8 \cdot 10^6$) TBq/Mton [2.5].

The maximum tritium inventory from nuclear explosions in the atmosphere was about $1 \cdot 10^8$ TBq [2.7] in 1963. Underground testing has not contributed significantly to the atmospheric tritium level [2.5]. The tritium inventory due to nuclear explosions is now about 25 times larger than the natural tritium inventory. Most of the tritium released by atmospheric testing enters the atmosphere as HTO. The oceans are a large sink for tritium.

An estimate of the amount of tritium in the reactor 3 at Tschernobyl yields values between 5.8 and 6.6 g [2.8]. Owing to the high temperatures during the accident, the entire tritium content was probably discharged. This corresponds to about 3% of the annual natural tritium production.

2.2 Risks Associated with Tritium

2.2.1 Incorporation

Due to the very low β^--decay energy of less than 18.6 keV, the effects on humans of external exposure to tritium are negligible. The β^- electrons will be absorbed in the outer layers of the skin. The maximum penetration depth of the β^- electrons in air is about 6 mm; in water and in the skin it is about 6 μm. On the other hand, the incorporation of tritium by inhalation, ingestion and submersion has to be considered. In addition, one has to distinguish between the toxicity of HT and HTO. A rule of thumb is that HTO is about 10^4 times more toxic than HT. Between 98 and 99% of inhaled HTO vapour is absorbed in body fluids, whereas in the case of HT or T_2 only 0.1-5% is absorbed in the lungs. A small part of the inhaled HT gas is oxidized to HTO in the body. Incorporated tritium will be more or less homogeneously distributed in the body with the exception of the skeleton. Therefore, a simple way to determine the tritium level in the body is to measure tritium in the urine with a liquid scintillation counter. Another method is to analyse the tritium concentration of the water vapour in the exhaled air. With these methods only the HTO component is measured. The biological half-life of tritium distributed as HTO in the body is about 8-12 days. These figures can be reduced to about 5-7 days by copious drinking. Beer is said to be the best fluid to clear the body of tritium. In addition, a small amount of tritium is organically bound (OBT) and has a biological half-life of 40-600 days. The OBT rises in the case of long term exposure.

The above discussion has shown that the radiation hazards from tritium depend strongly on the chemical form of the tritium contaminated compounds and on their ability to enter the human body.

A further interesting example is thymidine, which is used mainly in biological research. Incorporation of tritiated thymidine increases especially the risks of somatic and genetic damage, because thymidine is incorporated preferentially in the nuclei of cells. The maximum annual intake of tritiated thymidine should be smaller than that of tritiated water.

Further effects due to the incorporation of tritium may arise from the transmutation of tritium into ^3He, the recoil energy of the $^3He^+$ ion, the presence of ^3He in the body tissue after the β^- decay and the substitution of H by T before the β^- decay. Generally, these effects are negligible in comparison to the damage caused by the β^- electron. In addition, the radiation dose due to the generated antineutrino is negligible because the interaction of neutrinos and antineutrinos with matter is very weak.

So far, mainly the uptake of tritium from the surrounding air has been considered. Further paths may be direct skin contact with tritium contaminated surfaces, ingestion of tritium contaminated food and water and incorporation of metal tritide powder or tritium-labelled compounds. The last two materials result in poorly known radiation hazards due to the different retention and exchange times of tritium in these materials.

In the following, a few values will be given relating certain concentration values of T in urine and air to the effective equivalent dose. The values are taken from International Commission on Radiological Protection (ICRP 30) [2.9] and Strahlenschutzverordnung (StrlSchV) [2.10]. An averaged T concentration of 37 Bq/ml urine corresponds to an effective equivalent dose of 0.7 mSv/a (ICRP 30), or 5 mSv/a (StrlSchV) assuming HTO uptake. A dose of 0.01 Sv is obtained by a single incorporation of $9.6 \cdot 10^{12}$ Bq HT or $5.9 \cdot 10^{8}$ Bq HTO (ICRP 30), whereas a value of $8.9 \cdot 10^{7}$ Bq HTO is found by StrlSchV. Sievert (Sv) is the new SI unit for dose equivalent since 1986. 1 Sv corresponds to 100 rem.

Assuming a working time of 2000 h/a an equivalent dose of 0.05 Sv is obtained by the inhalation of air with a concentration of $2 \cdot 10^{10}$ Bq HT/m^3 (ICRP 30) or $1.2 \cdot 10^{6}$ Bq HTO/m^3 (ICRP 30), or of $1.78 \cdot 10^{5}$ Bq HTO/m^3 (StrlSchV). Considering also the permeation through the skin the last two of the above values have to be multiplied by the factor 2/3, because in the case of submersion 2/3 of the incorporation occurs through the lung and 1/3 through the skin.

In the following, a few maximum permissible concentrations of tritium in air or in drinking water, according to the German radiation protection ordinance (StrlSchV), are given. The maximum doses to the whole body for members of the general public and for radiation-exposed workers are, in general, 0.3 mSv (30mrem) and 50 mSv (5rem), respectively. The German radiation protection ordinance does not distinguish between HTO and HT intake. The whole-body dose limits are taken as the maximum permissible numbers for work with tritium. The derived maximum permissible values of the annual tritium incorporation, by inhalation or ingestion, are listed in Table 2.3 for members of the public and for radiation-exposed workers of Category A. With these values, the maximum permissible tritium concentrations averaged over one year in the air or in the water, released from radioactive laboratories, can easily be calculated, assuming a

Table 2.3. Maximum permissible values of the annual tritium incorporation by inhalation (air) and ingestion (food and water) for members of the public and radiation exposed workers derived in consideration of the 0.3 mSv (30mrem) or 50 mSv (5 rem) concept (for more details see text) according to StrlSch [2.10]

Maximum permissible tritium incorporation per annum for	Inhalation	Ingestion
Member of the public	$2.66 \cdot 10^{6}$ Bq/a ($72 \cdot 10^{-6}$ Ci/a)	$5.92 \cdot 10^{6}$ Bq/a ($160 \cdot 10^{-6}$ Ci/a)
Radiation exposed worker	$4.44 \cdot 10^{8}$ Bq/a ($12 \cdot 10^{-3}$ Ci/a)	$9.87 \cdot 10^{8}$ Bq/a ($26.7 \cdot 10^{-3}$ Ci/a)

Table 2.4. Maximum permissible values of the tritium concentration in air and water released from radioactive control areas and in the air in radioactive control areas according to StrlSch [2.10]

Maximum permissible values of tritium concentration averaged over one year in	air	water
Air or water released by radioactive areas	365 Bq/m^3 (9.86·10^{-9} Ci/m^3)	7.4·10^6 Bq/m^3 (2·10^{-4}) Ci/m^3)
Air in radioactive areas	1.78·10^5 Bq/m^3 (4.8·10^{-6} Ci/m^3)	-

breathing volume of 7300 m^3 air and a consumption of drinking water of 0.876 m^3 per year. These numbers are given in Table 2.4 together with the maximum permissible tritium concentration in the air of a radioactive laboratory. The higher maximum permissible tritium concentration in the air of radioactive laboratories is due to the 50 mSv (5rem) concept, and the shorter residence time of workers in controlled areas, which results in a maximum breathing volume of 2500 m^3/a. The values listed in Table 2.3 have to be observed. For comparison, the ICRP 30 recommended annual limits of tritium intake are 2.8·10^9 Bq (75·10^{-3}Ci) for ingestion as well as for inhalation in the case of radiation exposed workers: these values are about 4 times higher than those stipulated by the StrlSchV. The limit for free handling of tritium in Germany is 3.7 MBq (0.1mCi).

2.2.2 More Indirect Risks

Up to this point, only the direct contribution of tritium to the dose has been considered. In special cases, tritium can produce a radioactive dose indirectly. If tritium is absorbed in matter within the X-ray escape length, the energetic β^- electrons generated during the tritium decay can induce X-ray emission. These spectra are discussed in detail in Sect.4.3. Here the dose due to the X-rays, with their far larger penetration depth, has to be considered.

The chemical properties of tritium gas are very similar to those of D_2 and H_2 since these are determined mainly by the number of outer electrons and their arrangement. Due to the β^--decay energy, thermodynamic equilibria in certain compounds can be reached faster when tritium atoms are present. This means that in the case of tritium gas mixtures, the probability of explosions or detonations is even higher than for D_2 or H_2 mixtures.

Tritium gas can be transformed into tritiated water by oxidation (T_2+0.5O_2 → T_2O) and by isotope exchange (HT+H_2O → H_2+HTO). Both processes should be avoided from a radiation-protection standpoint, because tritiated water is much more dangerous than tritium gas. The rates of these reactions are slow at room temperature, but can be increased by the presence of catalysts, such as metal surfaces, ionising radiation, soil bacteria (in the soil the enzymes of microorganisms convert HT into tritiated water as the first reaction product), etc. The equilibrium constant of the exchange reaction given above is about 6.25 at 298K and 3.4 at 398K, showing that the heavier hydrogen isotope prefers to be built into the water molecule. This explains why most of the natural tritium on earth exists in the form of water.

Generally speaking, tritium belongs to the group of the less toxic radionuclides. This is mainly due to its short biological half-life in the body, to the low decay energy and to the low Linear Energy Transfer (LET).

An accident which led to an incorporation of about $3.7 \cdot 10^{10}$ Bq (1Ci) tritium, the subsequent treatment and the estimated dose, were discussed in a recent publication [2.11]. The tritium concentration in the body fluid of the workers in the Jülich 37 TBq (1000Ci) tritium laboratory is, on average, 0.5-1 Bq/ml ($1.3 \cdot 10^{-11}$ to $2.7 \cdot 10^{-11}$ Ci/ml). Until now, the highest concentrations in the urine were about 25 Bq/ml ($6.8 \cdot 10^{-10}$Ci/ml), which resulted from work with T-contaminated oil of rotary pumps.

2.3 Discharge of Tritium to the Environment

Besides the natural tritium production by cosmic rays, there are four important man-made sources of tritium discharge to the environment: 1) fission reactors and nuclear-fuel reprocessing plants; 2) tritium production plants; 3) consumer products containing tritium; and 4) nuclear weapons testing.

Values concerning tritium discharges during the year 1984 and the net contribution to the global environmental inventory at the end of 1984 are listed in Table 2.5 [2.3]. Almost no data are available for T production plants.

Table 2.5. Comparison of man-made tritium sources [2.3]

Source	discharge during 1984 [TBq/a]	total environmental inventory at the end of 1984 [TBq]
Nuclear fuel cycle	$\sim 1 \cdot 10^4$	$\sim 4 \cdot 10^4$
Consumer products	$\sim 10^4$	$\sim 10^5$
Nuclear explosions	-	$3.7 \cdot 10^7$

The total amount of tritium discharged into the environment by nuclear fuel cycles up to the end of 1984 is about 3% of the natural T inventory, and about 0.1% of the amount generated by nuclear weapons testing. As an example, a few values for the heavy-water reactors operated at Chalk River Nuclear Laboratories in Canada are [2.12]: total tritium inventory: $44 \cdot 10^3$ TBq (1.2MCi); mean HTO discharge to the atmosphere: 504 TBq/a (13600Ci/a), mean HTO discharge to the Ottawa River: 147 TBq/a (~4kCi/a). This yielded a concentration of 74 Bq/l ($2 \cdot 10^{-9}$Ci/l) at the plant boundaries in the atmosphere and a concentration of 10 Bq/l ($2.7 \cdot 10^{-10}$Ci/l) in the river at the position closest to a population site (compare Sect. 2.1.1). For the year 1984, the overall production of T generated by nuclear power stations was $1.1 \cdot 10^5$ TBq (3MCi) [2.3] of which 9% was released to the environment, which corresponds to about 14% of the natural annual production (Table 2.5).

Tritium is also released from T-containing consumer products, and these can be divided into 4 main groups [2.4]: 1) products with T-based luminous compounds containing 10^{-5} to 10^{-3} TBq ($2.7 \cdot 10^{-4}$ to $2.7 \cdot 10^{-2}$ Ci) apiece; 2) products with T_2-gas filled glass tubes coated with phosphor, each containing $4 \cdot 10^{-4}$ to 1 TBq (0.01 to 27Ci); 3) electron tubes, cold-cathode tubes, glow lamps, gas chromatographs, etc. containing $4 \cdot 10^{-8}$ to 10^{-2} TBq (10^{-6} to 0.27Ci) each; and 4) tritided n targets. The amount of tritium released from consumer products is about 8% of the natural inventory produced by cosmic rays (Table 2.5 and Sect.2.1.1). These man-made tritium amounts are expected to increase. In the consumer products mentioned above, the tritium decay is used as energy supply for ionisation and lighting in contact with special materials, e.g. phosphor. In high concentrations tritium itself is radioluminescent. The tritium decay product ^{3}He excited by the beta electrons contributes to the emission spectrum.

2.4 Tritium Handling in Future Fusion Reactors

Due to the complexity of future fusion reactor designs, only special components which are important for the production and handling of tritium will be discussed.

In a future power producing reactor, about 10 kg of tritium will have to be handled. This amount exceeds the total amount of natural tritium on earth (Sect.2.1.1) by a factor of three. Furthermore, the annual availability of T_2 gas will be only a few kg from CANadian Deuterium Uranium (CANDU) reactors, and probably no more than 5-10 kg/year from T_2-producing countries (United States, France, USSR) [2.13]. A 850 MW_t fusion reactor will require about $3 \cdot 10^{20}$ fusion reactions (2.2) per second, equivalent to a fuel burn up of about 3.6 g of deuterium and 5.4 g tritium per hour. This shows that only a very small amount of the total tritium inventory is needed to keep the plasma burning.

The major tritium inventories in future fusion reactors will be located in the following processing components: 1) plasma exhaust pumping, plasma exhaust impurity removal, plasma exhaust impurity processing, recycling; 2) blanket and blanket tritium recovery; and 3) isotope separation and fuel storage.

2.4.1 Multiple Containment

To avoid unacceptable environmental tritium release, vulnerable components have to be surrounded by secondary confinements with the smallest possible volumes, but allowing access for maintenance. The secondary confinements can be glove boxes containing the tritium processing equipment and detritiation systems or UHV chambers (Sect.3.4.4). Where the probability of accidental T release is high (as for cryopumps or cold traps [2.14]) a tertiary barrier [2.15] can be built in by isolating the ventilation of the volume with the T source and subsequent detritiation of the contaminated air. Thus, the best method for retention of T is the principle of multiple containment using materials with small T permeation coefficients [2.16] (Sects.3.4.5 and 3.7). With such multiple containments and detritiation systems, it is believed that the tritium discharges from future fusion reactors to the environment can be kept at values between 1.9 and 3.7 TBq/day. *Casini* et al. [2.17] estimated that the normal tritium releases from a 1000MWe fusion power reactor are about 777 TBq/a (21000Ci/a) consisting of 444 TBq (12000 Ci) HTO/a and 333 TBq (9000 Ci) HT/a. Comparing these numbers with those of Table 2.5 shows that just thirteen 1000 MWe power fusion reactors would already emit as much T as was discharged in 1984 in the whole nuclear fuel cycle of all fission reactors, which generated about 134 GWe·a [2.3] during 1984. This means that in a fusion reactor, 10 times more tritium will be released to the environment per MWe than in a fission reactor. On the other hand, these daily releases of T in a fusion power plant are lower than 1 ppm of the total tritium inventory and tritium is probably the only radionucleus released from a fusion plant during normal operation. The above-mentioned releases yield individual doses between 7.2 and 163 μSv per year for HTO and between $1.8\cdot10^{-4}$ and 3.4 μSv per year for HT at a distance of 1 km from the power station, according to different calculations made by *Edlung* [2.17].

2.4.2 Pumping and Recycling

Such important topics as plasma exhaust pumping, plasma exhaust impurity removal, isotope separation and breeding materials will now be addressed briefly.

Pumps and Impurity Removal

To keep the plasma burning, the fuel components deuterium and tritium have to be supplied continuously. Large pumping facilities must be installed because the fusion reaction product ^4He and other impurities such as CO, N_2, O_2, H_2, $N(D,T)_3$ and $C(D,T)_4$ have to be removed from the plasma. The pressure in the plasma exhaust device (diverter or pump limiter) will be about 10^{-4} mbar. Together with the ash He, unburnt deuterium and tritium will also be pumped. It is necessary to recycle the deuterium and tritium, and to recombine them in the proper proportions with tritium bred in the blanket. Pumps with high pumping speeds are cryosorption and turbomolecular pumps. A disadvantage of the cryopumps is that the tritium inventory may become very large and that they require periodic removal of the trapped gases. In the case of turbomolecular pumps, the tritium gas is in contact with the lubricating oil and may undergo isotope exchange reactions with the H atoms in the oil. As a consequence of the radiation dose, the viscosity of the oil and the life-time of the pump may change. Tritium compatibility measurements of turbomolecular pump oil (Leybold[1], type F12) have been performed with a miniaturised viscosimeter in the author's laboratory [2.18]. No viscosity change of the oil was observed during the test time which corresponds to about 250 years of operation of the JET tokamak with tritium.

A few general remarks about cryopumps are called for. All gases, with the exception of helium, cryocondense at 4 K; helium itself can only be cryosorbed at 4 K. Hydrogen gas mixtures and impurities can also be cryosorbed on molecular sieve or activated charcoal at 77 K. In addition to their excellent pumping properties, cryopumps can be used to separate more or less volatile gases. This is usually done by pumping the gas mixture with a cryocondensation pump and a cryosorption pump separated by valves. In the cryocondensation pump all gases are condensed, with the exception of He, which is only absorbed in the cryosorption pump. After closing the valves, the cryosorption pump can be heated to desorb the pure He gas. By distillation at 20 K the hydrogen gas can be separated from the other impurities in the cryocondensation pump. Tritium-compatible pumps were discussed by *Weichselgartner* [2.19] and *Hemmerich* [2.20], and vacuum applications for the tritium systems test assembly were presented in [2.21].

The tritium contaminated impurities are oxidised to water (HDO, HTO, H_2O, DTO, D_2O, T_2O), CO_2 and N_2. The water is collected by a cold trap and can be electrolysed to obtain pure hydrogen gas for recycling. Another method uses activated uranium which reacts with water forming uraniumoxide and uraniumhydride. Uranium reacts irreversibly with oxygen, nitrogen and carbon whereas hydrogen can be desorbed again at mod-

[1] Leybold-Heraeus GmbH: Standard Vacuum Components and Systems Division, Köln, FR Germany.

erate temperatures. A disadvantage of the last process is the generation of active solid waste.

Hydrogen Isotope Separation

Once the impurities of the plasma exhaust have been removed, it is necessary to separate H, D and T in the gas stream. One possibility is hydrogen isotope separation by cryogenic distillation. The difference in the boiling points of the gases H_2, HD, HT, D_2, DT and T_2 is used. The least (most) volatile gas is T_2 (H_2) with a boiling point of 25.04 (20.39) K. Thus, the more volatile isotope can be obtained at the head of the separation columns, whereas the less volatile isotope can be obtained at the bottom of the columns. Another technique is gas chromatographic separation [2.22] which (as mass spectrometry or laser Raman spectroscopy) can also be used for analysis of tritium gas mixtures. A disadvantage of this method is that it needs, in most cases, a carrier gas such as He, Ne or Ar and that it can separate only small amounts of mixtures. Further potential isotope separation methods are thermal diffusion [2.23] and a technique using resonance radiation [2.24] (see also Sect.2.5).

2.4.3 Tritium Breeding Materials

In the blanket which covers the first wall of the fusion reactor, the energy of the 14.1 MeV neutrons is deposited and transported to a conventional power plant. In addition, the fuel component tritium is also bred in the blanket (Table 2.2). Possible breeding materials under discussion are: 1) liquid lithium, which, due to its low melting point (186°C), can also be used as coolant and effectively thermalizes the neutrons. The T bred is transported with the liquid Li stream and can be separated externally; 2) lithium-lead alloys such as $Li_{17}Pb_{83}$ with a melting point at 235°C and a high breeding rate due to the neutron multiplying element lead (Sect.2.1.3). A disadvantage of $Li_{17}Pb_{83}$ is the extremely low solubility of T in it, which leads to a high permeation rate through the hot stainless steel walls; 3) ceramics such as lithium oxide (Li_2O, LiO_2), lithium aluminate ($LiAlO_3$, $LiAlO_2$) and Li compounds containing silicon (Li_2SiO_3, Li_4SiO_4) and zirconium (Li_4ZrO_4, Li_2ZrO_3), which only breed enough T if neutron multipliers are used. The tritium outgassing from the ceramic compounds will be transported from the blanket in a He stream and must be recovered externally. The materials mentioned above have been studied extensively [2.25-37], because tritium production is an essential prerequisite for future fusion reactors.

2.4.4 Planned Tritium Use at JET

At the Joint European Torus (JET) experiment the first use of tritium is planned in the years 1991-1992. The tritium activity employed per pulse will be 90 TBq (≃2450Ci) [2.38]. The recycling system will be able to reprocess about 48 pulses per day. The total tritium content of the recycling system will be about 10 g (≃10^5Ci) [2.38]. This number was recently increased to 90 g (≃$9 \cdot 10^5$Ci). During the operation phase only tritium amounts up to two-thirds of it will be handled in gaseous form [2.39]. The tritium content in the INternational TOkamak Reactor (INTOR) is estimated at 3.3-5.7 kg [2.17], whereas the corresponding numbers for the Next European Torus (NET) are 2.2-3.6 kg [2.15].

As long as graphite is used as first-wall material, a part of the tritium fed to the tokamak will be absorbed by the graphite. Estimates show that the tritium activity accumulated after 1000 shots in the JET torus will be about 890 TBq (24kCi) [2.40].

An accidental release of 10 g (≃3700TBq or 10^5Ci) tritium in the form of tritiated water at ground level at JET would result in an estimated dose of 30 mSv (3rem) for an individual at the plant boundary, due to direct intake of airborne tritium [2.41]. The probability of such an event is estimated to be less than $1 \cdot 10^{-4}$ per year. Another accident scenario, where the whole tritium inventory of JET would be discharged to the river Thames, would lead to a dose to an individual of 0.3 mSv [2.41], if the water is used as drinking water.

Due to the 14 MeV neutron of the D-T reaction, the structural components of a fusion reactor will be radioactive. However, this activity does not lead to exposure of the public during normal operation. Furthermore, major explosions similar to nuclear explosions cannot occcur in fusion reactors, because the total amount of reacting fuel in the plasma is unable to maintain full power operation for more than a few tens of seconds, and any disturbance leads to loss of plasma confinement and hence terminates the fusion reaction. The main advantage of fusion reactors is that the fuel components deuterium and lithium, and the final reaction product ^{4}He, are not radioactive substances and that the intermediate fuel tritium can be handled in a more-or-less closed loop.

Small fusion device fuel clean-up systems have been described in [2.42-44].

2.5 Tritium Waste Handling

Tritium waste handling has been discussed in detail previously [2.7, 45-53]. In this section only a few aspects will be briefly mentioned.

Because tritium in the chemical compounds HT or HTO may be easily dispersed worldwide, tritium waste generated in discarded tritium containing consumer products (Sect.2.3) [2.54-56], conventional nuclear reactors and future fusion plants must be carefully handled to avoid environmental contamination. A CANDU reactor, for example, contains up to 1 TBq (~30Ci) of tritium per kilogramme of heavy water in the moderator, which corresponds to a ratio of DTO : D_2O of twenty parts per million. At present, nuclear reprocessing plants discharge half of the tritium produced as HTO into the sea (e.g., at La Hague in France, and Sellafield in the UK) or as HT and/or HTO into the atmosphere (e.g., USA and India) [2.48-50]. The remainder is incorporated as zirconium tritide in "Zircaloy" material used as cladding; (about 15m^3 cladding waste is produced per $GW_e a$). Zircaloy is a zirconium based alloy.

The increasing amount of man-made tritium from consumer products and nuclear reactors requires more research into the techniques of tritium disposal and recycling. Tritium waste may be categorised according to its chemical bonding as aqueous, organic and elemental (gaseous) waste.

Some possible disposal methods are as follows: All three categories of waste may be stored in double-contained stainless steel vessels; the maximum tritium concentration recommended by ICRP should not exceed 3.7 TBq (~100Ci) per kilogramme of water. Aqueous wastes may be bonded in concrete, poured into drums or barrels, (for storage or eventual deep sea disposal), fixed "in situ" by concrete poured into salt domes, or injected into deep geological formations. Tritium gas may conveniently be stored in metal tritides: zirconium, titanium, and rare earth elements are eminently suitable for this purpose. Leaching rates for metal tritides are orders of magnitude smaller than those of cement. All the techniques mentioned above require further development.

The main hazard of stored tritium waste is leakage. Containers must be able to withstand both shock and corrosion damage as well as the increase of internal pressure resulting from the decay of tritium to helium-3 over a period of at least a century. The choice of disposal method depends upon site-specific data such as geology, hydrology, meteorology and the possible local effects of tritium leakage. Immobilisation must be both reliable and publically acceptable.

High tritium concentrates, such as those produced by heavy-water reactors in the moderator, may be reduced by extraction and enrichment of the tritium content using special techniques such as water distillation, electrolysis, liquid- or vapour-phase catalytic exchange, or laser pre-enrichment in conjunction with cryodistillation. In water distillation, the tritium-containing molecules HTO and DTO become concentrated in the liquid as they are removed from the vapour, with the advantage that the handling of explosive gas mixtures (H_2, HT, D_2) is avoided. Disadvantages are the enrichment of tritium in its more radiotoxic liquid form, and the combination of high energy consumption with small separation factor of this process. In electrolysis, the electrolyte water becomes enriched with tritium at the expense of the tritium in the hydrogen gas formed at the cathode. For the recovery of tritium, the main purpose of electrolysis is to decompose water. The isotopic separation of the hydrogen molecules is done in a subsequent cryogenic distillation system or a gas chromatography system.

The catalytic exchange process with subsequent cryogenic distillation is characterised by the following reactions (X: H or D):

a) chemical exchange $XTO + X_2 \rightarrow X_2O + XT$,

b) cryogenic distillation $2XT \rightarrow X_2 + T_2$.

The products of these processes are detritiated water and isotopically pure gases, which may then be used for other purposes.

A plant using vapour-phase catalytic exchange in conjunction with cryodistillation to detritiate heavy water has been in service at ILL in Grenoble since 1973 [2.57]. Two more such plants are being built in Canada (at Chalk River and a larger one at Darlington) with liquid- and vapour-phase catalytic exchange, respectively, to detritiate heavy water from the CANDU reactors mentioned previously.

In the case of elemental tritium waste, the most attractive method is to separate tritium gas from gas mixtures, thereby reducing the waste volume and recycling the tritium. Well-known techniques to remove tritium from gas mixtures are cryogenic distillation, and on a smaller scale, gas chromatography, and thermal diffusion. A safe method for the storage of tritium gas is the use of metallic getter materials which form stable metal tritides with very low equilibrium pressures at room temperature (Chaps.3 and 5).

The different separation and/or enrichment methods of tritium mentioned above were discussed in a recent book by *Vasaru* [2.58].

3. Preparation of Metal Tritides

Metal tritides can be produced by similar physical procedures and techniques as metal protides and metal deuterides with necessary modifications to ensure safe tritium containment. The following tritium-charging techniques will be discussed briefly: 1) electrolytic charging, 2) irradiation in nuclear reactors, 3) production in accelerators, 4) charging from the gas phase, 5) charging of metals using metal tritides.

3.1 Electrolytical Charging of Metals with Tritium

The electrochemical doping [3.1] of cathodically polarized metals in a liquid electrolyte is a frequently used procedure, due to the simple experimental setup required. Owing to the presence of atomic hydrogen in the electrolyte-metal interface, most metals can easily be charged with hydrogen up to large concentrations. Examples of electrolytes employed for the production of MH_r and MD_r samples are HCl, H_2SO_4, HNO_3 and DCl, D_2SO_4, DNO_3, etc., respectively. In the case of pure tritium, the protons or deuterons have to be exchanged for tritons. As far as is known, no electrochemical doping techniques using 100% tritiated acids (TCl, T_2SO_4, TNO_3) are in use. The main disadvantages of such a procedure would be: 1) superheavy water vapour, which is (Sect.2.2) about 10000 times as hazardous as tritium gas, has to be dealt with; 2) pure tritium gas is generated at the cathode; and 3) the activity contained in 1 g superheavy water is about 97.4 TBq (2633Ci). For the electrochemical doping of metals with tritium, only acids with low tritium concentrations have been employed to date (for example: 74 GBq (2Ci)/ml in [3.2]). Samples charged in such a way contain a mixture of H or D and T atoms and have been studied mainly in tracer and autoradiography experiments.

3.2 Breeding of Tritium in Metals

Tritium can be bred by the capture of neutrons in nuclear fission reactors, as mentioned in Sect.2.1.2. By means of this procedure, large samples can be charged homogeneously with tritium. The main disadvantages of this technique are: 1) the method works only for a very limited number of isotopes; 2) the other isotopes of the sample can be activated resulting in highly radioactive samples; 3) radiation damage occurs in the sample, which complicates further studies and interpretations; and 4) the high costs.

3.3 Tritium Charging of Metals by Means of Nuclear Reactions and Ion Implantation

Metals can be loaded with hydrogen or deuterium by ion implantation in accelerators. In the case of the stable hydrogen isotopes, very high local hydrogen concentrations have been obtained which were difficult to achieve by electrolytical or gas-phase charging. Only very few accelerators exist for implantation of tritium in metals[1]. Obviously, in such experiments all the inner surfaces of the accelerator become contaminated with tritium.

For producing tritides, accelerators can be employed using nuclear reactions such as (p, t), (α, t) and others producing tritons. All metals can be doped in this way if a material undergoing such a nuclear reaction is bombarded just in front of the metal or alloy to be charged with tritium. If the tritons produced by the nuclear reaction have sufficient kinetic energies to leave the first material, they will be implanted in the metal to be charged [3.4]. The main disadvantage of this technique is again the introduction of lattice defects and inhomogeneous doping of large samples.

3.4 Tritium Charging of Metals from the Gas Phase

Metal tritides are usually produced from the gas phase. In this way, the absorbed amount of tritium in a solid can be varied by changing the sample temperature and the tritium pressure. This procedure is also used in the $3.7 \cdot 10^{13}$ Bq (1000Ci) tritium charging apparatus [3.5,6] in the author's laboratory, which is very similar to that used to prepare MH_r and MD_r samples from the gas phase [3.1,7]. It was necessary only to build in the modifications required to ensure safe handling of tritium. These modifications were: the construction of an all-metal ultrahigh-vacuum (UHV) apparatus and the construction of secondary containments which surround hot walls in contact with high tritium pressures, in order to capture tritium that has permeated through the hot walls. In addition, the whole tritium charging equipment is placed in a ventilated hood in a controlled radioactive laboratory, and the tritium level in the exhaust air from the hood is continuously monitored.

---
[1] Eg, Chalk River 70 keV isotope separator equipped for handling large tritium quantities [33].

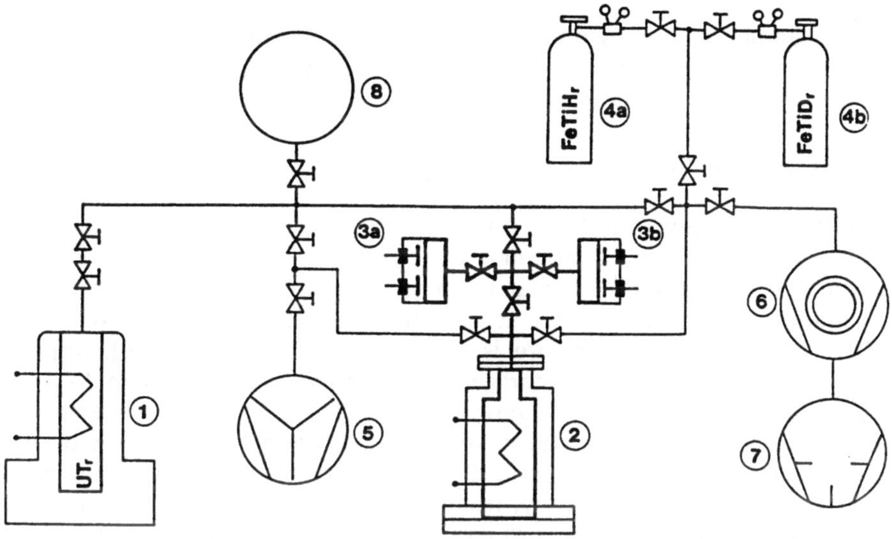

Fig.3.1. Schematic picture of a 37 TBq (1000 Ci) tritium charging apparatus [3.5]. (1: UT_r storage container, 2: sample chamber, 3a: single-sided capacitance manometer (full range: 1.33 mbar), 3b: single-sided capacitance manometer (full range 13.33 bar), 4a: $FeTiH_r$ storage container, 4b: $FeTiD_r$ storage container, 5: ion getter pump, 6: turbo-molecular pump, 7: rotary pump, 8: additional volume)

The activity of 1 cm^3 T_2 gas at standard pressure (1atm $\hat{=}$ 1.013bar) and temperature (273K) is $9.6 \cdot 10^{10}$ Bq (2.59Ci). The price of tritium gas varies between 3 and 15 DM per $3.7 \cdot 10^{10}$ Bq (1Ci) for low quantities. Tritium gas of purity better than 98% is commercially available. Tritium gas up to $3.7 \cdot 10^{12}$ Bq (100Ci) can be bought in glass ampoules which are readily shipped.

A scheme of the author's all-metal 37 TBq (1000Ci$\simeq$0.1gT) tritium loading apparatus is shown in Fig.3.1. The main components, the uranium tritide storage container (*1*), the sample chamber (*2*) and the capacitance manometers (*3a* and *b*) will be described in more detail below. The iron titanium protide (*4a*) and deuteride (*4b*) storage containers [3.8,9] deliver the H_2 and D_2 gas for isotopic studies of the solubility. After a sample change, the air in the sample chamber is pumped by a turbo-molecular pump (*6*) and a rotary pump (*7*). The final vacuum is obtained with the ion getter pump (*5*) which is also used to pump the residual noble gas not absorbed in the UT_r storage container (*1*). Two valves in series ensure that the tritium is safely contained and prevent it (which during solubility measurements fills only the small volumes indicated by the thick lines in Fig.3.1) from reaching the hydrogen storage containers and the pumps by an erroneous operation of a valve.

To date, no increase in the count rates of the tritium monitors has been observed when heating the UT_r storage container and/or the sample chamber. Only a small increase of tritium inside the hood is measured during sample exchanges due to the desorption of tritium from the inside chamber walls. The amount of tritium released during one opening of the sample chamber is in the range of 3.7 MBq ($10^2\,\mu$Ci) to 37 MBq (1 mCi).

3.4.1 Uranium Tritide Storage Containers

Uranium is employed as the tritium storage material (see also Sect.3.6.1). The main advantages of uranium are: 1) Very low tritium equilibrium pressures of about $2\cdot10^{-8}$ bar at room temperature [3.10]; this means that the uranium works as a getter at low temperatures. 2) At about 400° and 550°C the equilibrium pressure is about 1.2 and 14.1 bar [3.10], respectively. Therefore, the UT_r storage container can release tritium (T_2) at high pressures when the UT_r powder is heated up to moderate temperatures. 3) Uranium disintegrates to very fine powder after reaction with hydrogen. Due to this pulverization, the surface area capable of reacting with hydrogen increases dramatically, resulting in very short absorption times. 4) The uranium powder also reacts easily with oxygen, nitrogen and hydrocarbons in the temperature range employed and, therefore, works as a gas purification system. Disadvantages of the use of uranium are: 1) its radioactivity, 2) the powder is highly pyrophoric when exposed to air, and 3) the spread of the powder has to be avoided.

A cross-section of a UT_r storage container built in the author's laboratory [3.11] for $3.7\cdot10^2$ TBq (10^4Ci) tritium is shown in Fig.3.2. Multiple containment was used for the safe enclosure of tritium and UT_r powder. The inner volume (*3*) containing the UT_r powder is surrounded by two further volumes (*2* and *1*) which can retain the tritium in case of any leakage. For the decomposition of the UT_r powder, the oven (*7*) in the second containment (*2*) is heated. During this process, the inner walls are also heated and some permeation of tritium through the hot walls into the argon filled volume (*2*) occurs. To exclude any permeation of tritium into the biosphere, the volume in the centre (*2*) is surrounded by a tertiary containment (*1*) which is filled with circulating water of a closed primary cooling loop. The water is checked for tritium twice a year and exchanged.

The uranium (*5*) is contained in quartz or stainless steel cylinders (*6*) to avoid alloying of the uranium with the first containment (*3*) at temperatures above 630°C. About 80 g of uranium (*5*) are used in two chambers. Stainless steel filters (*4*) with a mesh size of 5 μm separate the chambers and prevent the transport of the UT_r powder from the UT_r storage volumes.

The temperature of the oven (*7*) is measured by two shielded thermocouples (*8*). The current feed-throughs (*11*) are protected with a vacuum-sealed cylinder. The top UHV flange (*10*) is shielded with a water-cooled Cu plate (*9*). Nupro[2] UHV valves (*12*) are directly welded to the UT_r storage container and the top flange.

---

[2] Nupro Company, 4800 East 345. Street, Willoughby, Ohio 44094, USA

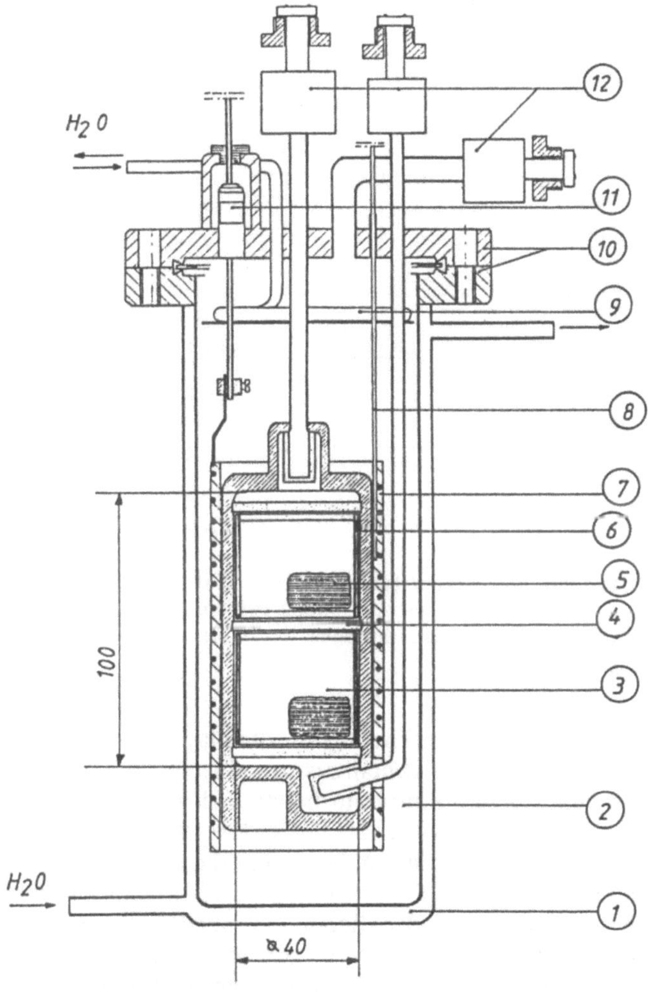

Fig.3.2. Cross-section of a 370 TBq (10000 Ci) UT_r storage container [3.11]. (1: water-cooled tertiary containment, 2: secondary containment filled with argon at 0.5 bar, 3: primary containment, 4: stainless steel filters, 5: solid uranium before hydriding, 6: quartz or stainless steel cylinder, 7: oven, 8: shielded thermocouples, 9: water-cooled Cu shield, 10: UHV CF flange, 11: current feed-through, 12: valves)

3.4.2 A Typical Sample Chamber

A cross-section of the sample chamber used to prepare metal tritides for experiments outside the charging equipment and for solubility measurements is shown in Fig.3.3 [3.11]. Again the construction follows the principle of multiple containment. The oven (*8*) is located in the sample chamber (*1*) so that the walls (*7*) can be cooled by the circulating water of a closed loop to reduce permeation of tritium. Permeating tritium is captured in the

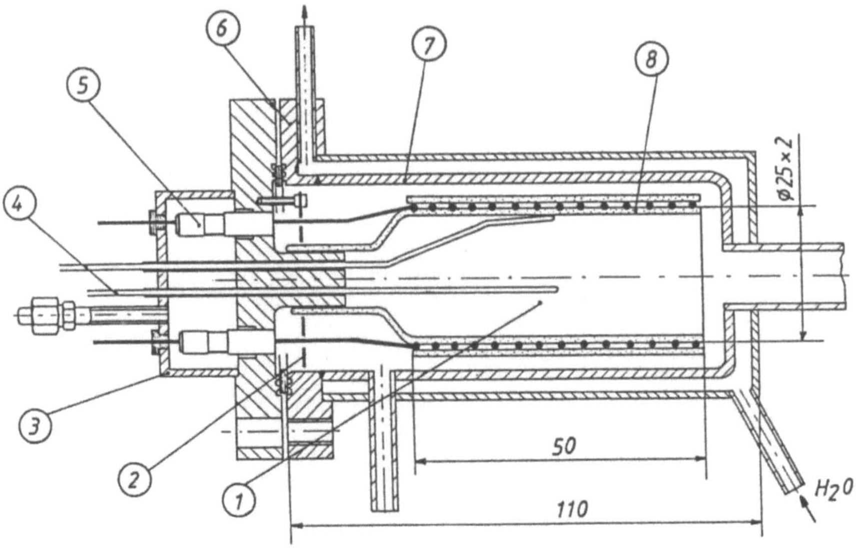

Fig.3.3. Cross-section of the sample chamber for preparation of MT_r samples [3.11]. (1: primary containment, 2: heat shield, 3: vacuum sealed cylinder filled with argon gas, 4: shielded thermocouples, 5: current feed-throughs, 6: double-knife-edge flange, 7: water-cooled inner walls, 8: quartz oven)

water which is replaced twice a year. The main advantages of this construction in comparison with the UT_r containers are: 1) the possibility of very fast temperature changes of the sample to be studied and 2) very low permeation rates due to the cooled walls of the first containment. Two shielded thermocouples (*4*) are used - one to determine the sample temperature and one to regulate the power to the oven. The current feed-throughs (*5*) are protected with a vacuum sealed cylinder (*3*) filled with argon to prevent any tritium release in the event of a rupture in one of the current feed-throughs. Argon prevents the build-up of dangerous and explosive gas mixtures. The cylinder (*3*) was filled with argon rather than with helium, to allow further use of highly sensitive He leak detectors even in the event that a leak should occur between the volume inside the cylinder (*3*) and the sample chamber (*1*). The whole oven is built on a 2 3/4" conflat flange (*6*) protected by a heat shield (*2*) from heat radiation and convection.

3.4.3 Capacitance Manometers

This section demonstrates that pressure gauges which work well for nonradioactive gases can give erroneous results with radioactive gases. Single- and double-sided capacitance manometer heads will be compared. Sensor heads of the double-sided version have electrodes on both sides of the pressure sensing diaphragm, whereas, for single-sided versions, both elec-

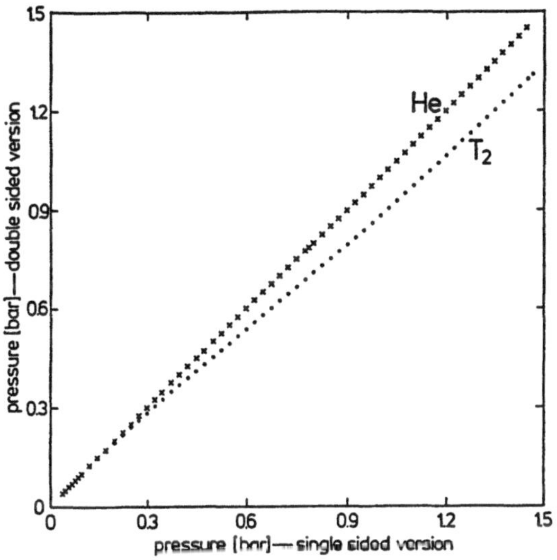

Fig.3.4. T_2 and He gas pressures measured with the double-sided capacitance manometer as function of the pressure determined with a single-sided version [3.5, 12]

trodes are on the reference vacuum side. The displacement of the diaphragm with pressure changes the capacitance which is measured with 10 ppm resolution. If the dielectric constant of a gas is constant, the output voltage of a capacitance head is directly proportional to the applied gas pressure. Figure 3.4 shows the absolute pressures for pure He and T_2 gas obtained with single- and double-sided capacitance heads. The T_2 pressure values measured with the double-sided version are lower than the He values, which are in very good agreement for both types. The data are presented differently in Fig.3.5 where, on a double logarithmic scale, the relative pressure difference between the single- and double-sided version is plotted as a function of the absolute pressure obtained with the single-sided version. The dashed line shows the systematic error, as stated by the manufacturers. In the case of He, the two pressure heads agree very well, illustrating their equivalence for nonradioactive gases. The somewhat higher error of the data in the pressure range below 60 mbar is probably due to the high sensitivity of capacitance gauges to vibrations induced by pumps. Below about 50 mbar, the two types of pressure gauges also yield satisfactory agreement in the case of T_2 gas. However, above 50 mbar the relative error in the pressure difference increases to about 12% at about 1 bar T_2 pressure. This deficiency of the double-sided capacitance head is caused by the considerable conductivity of T_2 gas due to the β^- decay of the tritium atoms (1.1) within the sensor head. This conductivity - although very disturbing in the case of the double-sided capacitance head - was used to construct a very simple, cheap and UHV-compatible tritium pressure gauge in the range from 10^{-3} to 10 bar which is also bakeable to

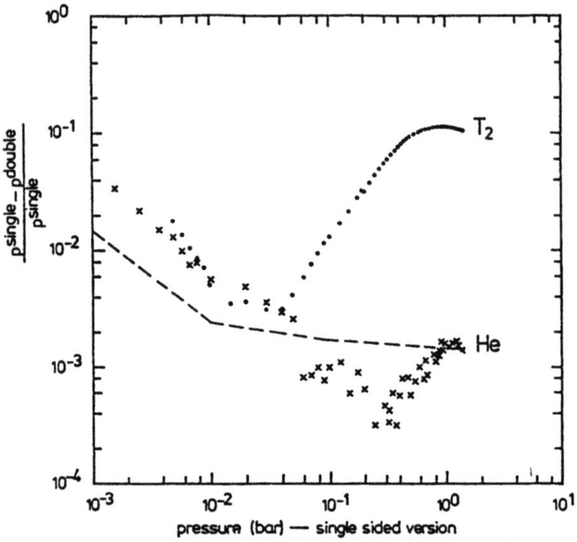

Fig.3.5. The difference between the pressures measured with capacitance manometers of the single- and double-sided version relative to the pressure of the single-sided version as function of the pressure obtained with a single-sided head for T_2 (•) and ^4He (x) [3.5, 12]. The dashed line represents the sum of the systematic errors given by the manufacturers

250°C and insensitive to vibrations [3.12]. Thus, to measure pressures of radioactive gases, the use of double-sided capacitance heads should be avoided [3.12-15].

3.4.4 Glove Boxes

As mentioned in Sect.2.4, the vulnerable parts of any tritium storage and processing equipment have to be surrounded by secondary confinements in order to avoid tritium release into the environment. Special glove boxes built in such a way that maintenance and repair of the first containment can easily be performed are often used for the secondary containment. Access to the tritium-containing apparatus (first containment) inside the glove boxes should be possible through ports fitted with gloves which possess a good wear resistance and low permeability to water. Glove port covers should be employed when the gloves are not used. The glove-box gloves should only be touched with hands covered by easily disposable gloves to reduce tritium uptake through the skin. Information about hydrogen isotope permeation through elastomeric materials can be found in [3.16-18]. These glove boxes should also contain an integrated tritium clean-up system which is able to remove any tritium that leaks through the first containment or is released during an accident. Materials are moved into and out of the glove box by a transfer port. The tritium-contaminated gas in the transfer port can be exchanged prior to opening. Penetrations through the wall of the glove box are provided for process and service tubes, instrumentation and power. The overall leak rate should be less than 0.25 vol%/h. Glove boxes filled with inert gas are often operated at pressures slightly higher than ambient to reduce ingress of air. A brief comparison of potential tritium recovery techniques for use with helium was given in [3.19].

The physical principle of most tritium clean-up systems used in glove boxes is based on the catalytic oxidation of tritium and tritiated compounds and the subsequent adsorption of the tritiated water on molecular sieves. Pd-based catalysts at low temperature (100 to 200°C) or copper oxide at higher temperature (400 to 600°C) can be used for the oxidation. The generated tritiated water is adsorbed on molecular sieves or other water absorbents with very low water vapour pressure at room temperature. A very low tritium concentration is achievable with molecular sieves; about 1.85 MBq/m^3 ($5 \cdot 10^{-5} Ci/m^3$) is obtained for a waterload of 0.1% at room temperature. This means that voluminous containers filled with molecular sieves must be employed to avoid frequent change or reactivation of the molecular sieves. The regeneration of molecular sieves with tritiated water inventory was discussed in [3.20].

Glove box atmospheres should be inert to avoid explosive gas mixtures during large tritium discharges. However, in such glove boxes, oxygen has to be introduced for the oxidation in the clean-up system. From a radiation protection standpoint, the oxidation of tritium gas to tritiated water should be avoided because the aqueous form is about 10^4 times more toxic than the elemental form (Sect.2.2). Tritium clean-up systems which getter tritium gas consist of an inert gas purifier containing hot metal beds with high affinity for oxygen, nitrogen, water and hydrogen. The inert gas is circulated through a hot titanium bed which removes oxygen, nitrogen and water vapour, while a second bed at lower temperatures removes the hydrogen isotopes [3.21]. This method has the advantage that no low-temperature systems are required. Disadvantages are that at high temperature, tritium permeation losses can occur during regeneration, and solid radioactive waste is generated.

Other tritium removal systems use the catalytic hydrogenation of liquid organic compounds, for example linolic or linolenic acid [3.22]. The tritium atoms are strongly bound to the unsaturated molecules of these fatty acids. Until now, this method has only been applied to glove boxes filled with inert gases.

UHV chambers containing the tritium apparatus can also be used as secondary containment. These can be either evacuated or filled with inert gas at subatmospheric pressure. Any leaks through the first and/or second containment cause an increase of the pressure inside the UHV vessel. Tritium leaks are also detected by changes of the signal of the built-in ionization chambers. If there are no leaks, the pressure will become constant after a certain time when the desorption and adsorption rates of the gas atoms at the inner vessel surfaces are equal. Before the UHV vessel can be opened, the tritium in the first containment must be stored in a safe way and the volume of the second containment purged to transport the potentially tritium-contaminated gas to an external tritium removal system. For repairs to the first containment, gloves have to be installed in the second containment to allow safe access to the first containment.

3.4.5 Material for the First Containment

The materials for tritium handling facilities must be chosen very carefully. The important parameters, which influence the choice of the most suitable materials include: solubility, diffusivity, permeability, outgassing behaviour, bakeability, strength, corrosion behaviour, void swelling, fatigue, hydrogen embrittlement and stability under radiation, each as a function of temperature (Chap.5), and the ^{3}He behaviour (Chap.6). *Maroni* and *Deventer* [3.23] partly addressed these properties for many elements and alloys.

Probably the most frequently used construction materials for tritium handling facilities are 316L and 304 stainless steel, which are similar to the European stainless steels 1.4404 and 1.4301, respectively. 400 series stainless steel and inconel should be avoided. At NET the following two materials options for the first wall and breeder structural components are discussed: 1) solution annealed or cold worked 316L austenitic steel and 2) quenched and tempered 1.4914 martensitic steel. The integrity of austenitic stainless steel for tritium storage was treated by *Ells* [3.24]. In older apparatus, the manifold and tubing were often made of copper. Copper has the advantage of lower tritium permeability in comparison to stainless steel, but is for many applications too soft. In the past, most stainless steel tube connections were welded, with the disadvantage that, in the case of a repair, hot tubes had to be cut and welded again.

Nowadays, due to the fact that UHV couplings are commercially available, for example VCR[3] couplings, parts which have to be exchanged after a certain time are connected by these demountable couplings to the tubing for easy removal. Further, well suitable couplings are: CF-, KF-type flanges, helicoflex gaskets, etc. The integral tritium leak rate of an apparatus built with these UHV couplings is about the same as one which is totally welded. The integral leak rates of such tritium handling apparatus should be as low as possible. With commercial ^{4}He leak detectors, leaks of the order of 10^{-9} mbar·ℓ·s^{-1} can be detected easily. Such a tritium leak corresponds to a permeated tritium amount of about 84 Bq/s ($2.3 \cdot 10^{-9}$ Ci/s or ~0.2 mCi/d). Tritium leaks can also be detected using a sensitive tritium monitor for sniffing along the tubing. Useful tips on tritium technology are presented by *Anderson* et al. [3.25]

[3] The VCR coupling is a special type of bakeable UHV connector using metal gaskets giving a leak rate of less than 10^{-9} mbar·ℓ·s^{-1}, while providing high pressure capability. It has been developed by the CAJON Company, 9760 Shepard Road, Macedonia, Ohio 44056.

The treatment of the inner surfaces of the first containment as well as its cleanliness are very important prerequisites for maintaining the tritium gas purity [3.26]. Smooth interior surfaces are highly recommended. It is well known that tritium is able to react with carbon in stainless steel. Tritiated methan is produced. The concentration of methan at some time is proportional to the partial pressure of tritium and the surface-to-volume ratio of the container. The decrease of the tritium gas purity can also be reduced by the use of other clean materials, e.g. glass, Al, Cu, etc., instead of stainless steel.

In the primary containment, the use of organic polymers for seals and valve seals must be avoided, because they degrade under radiation in the presence of concentrated tritium gas. Furthermore, the use of oil lubricated pumps should be avoided, because tritium atoms will exchange with the protium atoms of the oil and, as a consequence, the oil properties may change. This means that elastomers (e.g., viton), halogenated materials (e.g., teflon) and organic lubricants (e.g., oil) should never be used in tritium process lines.

Valves, pressure gauges, flowmeters, etc., totally made of metal are supplied by different companies nowadays. Also oil-free pumps are commercially available, e.g. metal bellow pump, scroll pump, wobble pump, adsorption pump, getter ion pump, etc.

The best primary containment is a carefully planned and built all-metal construction. Each material should be checked for its tritium compatibility. Stringent quality assurance measures should be taken during construction, e.g. repeated material controls, ^{4}He leak tests, pressure tests, radiographing of all welds, etc.

3.5 Tritium Charging of Metals by Means of Metal Tritides

The surfaces of most metals are covered with oxide layers which prevent the absorption or desorption of hydrogen. This is caused by a dramatic decrease of the dissociation probability of X_2 ($X = H, D, T$) molecules or of the recombination probability of X ($X = H, D, T$) atoms due to the reduction of catalytic sites in the presence of oxygen atoms at the surface. To circumvent this problem, most metals have to be activated, which means they have to be temperature cycled in high vacuum or hydrogen gas. These temperature cycles cause cracks in the oxide layers, dissolution of oxygen atoms in the interior of samples, and reactions to water. As the main result, the number of oxygen atoms at the surfaces decreases, enabling more catalytic metal atoms to be at the surface to start absorption or desorption of hydrogen.

A disadvantage of this activation process is the increased quantity of oxygen amount in the metals which, in many cases, cannot be tolerated due to the trapping of dissolved hydrogen atoms near oxygen atoms. This can be avoided by contacting a hydrogen-loaded metal sample with the metal to be charged. Very good contacts can be generated using spot welding machines. The hydrogen diffuses through the generated welding points from the hydrogen doped sample into the sample to be charged. During the spot welding process only a very small part of the sample will be heated, and the oxygen atoms can only diffuse during the very short time of welding. The main part of the samples thus loaded (mainly foils) will therefore be free of oxygen contamination. The charging times depend on the number and quality of the junctions, on the dimensions of the foils, and on the temperature.

This technique [3.27] has been used to charge tantalum foils with tritium to determine the diffusion of tritium in Ta at very low temperatures using the perturbed angular correlation (PAC) method, because trapping of T atoms by oxygen atoms had to be avoided. In the case of tritium, the spot welding parameters (power, rise time, decay time, duration of pulse, etc.) should be optimised, using similar protided or deuterided metal samples, to avoid tritium releases in cases of wrongly chosen parameters.

This method can also be used to extract tritium from a metal tritide at room temperature by spot welding it on a pure metal with a higher affinity to tritium.

3.6 Metals Employed for Tritium Storage

Before discussion of the metal tritides typically used in tritium handling facilities and for tritium waste disposal, a few advantages of the storage of tritium in metals are discussed.

A very important advantage of storage of tritium in metals is the small dimensions of the metal tritide storage containers (Fig.3.2) in comparison to storage of tritium gas in the voluminous containers which are made necessary because tritium gas should be stored only at a pressure below 1 bar (a). As an example, 100 ℓ tritium gas at a pressure of 0.9 bar can be stored in ~80 cm^3 as LaT_2. The metal tritide storage container should be so constructed that heating of the metal tritide is impossible with valves closed, because very high pressures are generated in the small container volume upon heating. The storage of tritium in metal tritides possessing a low equilibrium pressure avoids excessive reaction of the tritium in the gas phase with the container walls, thus reducing the deterioration of the tritium purity in comparison to the storage of T_2 gas. Furthermore, a low equilibrium pressure reduces the permeation of tritium through the container walls into the environment. Due to the very small solubility of the tritium decay product, ^3He, in metals, ^3He can easily be separated by temperature cycling the metal tritide because upon cooling only the tritium gas will be absorbed. The unabsorbed ^3He gas has only to be removed from the cold metal tritide storage container. (In this way ^3He gas can be collected because it has a significant commercial value (150DM/g). Natural gas wells can contain ^4He concentrations up to several percents. ^3He concentrations are far below 1 ppm making the extraction of ^3He too expensive. Thus the decay of tritium is a very practical source of ^3He). In tritium gas storage containers such a simple separation of the tritium molecules from the helium atoms is not possible. With the choice of suitable metal tritides, the use of mechanical pumps and compressors, which are vulnerable parts of a tritium facility needing regular maintenance, can be avoided. The only mechanically moving parts left are valves.

Most metals have to be activated to obtain high absorption and desorption rates. This procedure involves temperature cycling the metals in the presence of hydrogen gas. At higher temperatures part of the oxygen in the oxides at the surface, which prevent the dissolution of the hydrogen in the bulk of the metal at low temperatures, diffuses into the metals, leaving catalytically active metal atoms at the surface which then enable the hydrogen molecules to dissociate and the hydrogen atoms to be absorbed into the lattice. Metals that are to be used for the storage of very pure tritium gas should be activated only with deuterium because this is a less disturbing impurity than protium from an isotopic standpoint. The activation with deuterium instead of protium also helps to eliminate the slight protium contamination present in most virgin materials from the fabrication process.

A disadvantage connected with any tritium absorbing material is the blanketing effect on the absorber of the tritium decay product, ^3He, which is always present in tritium gas and may prevent the tritium molecules from reaching the absorber surface, because even at an initially low He concentration, the relative amount of ^3He in the gas increases during the T_2 absorption forcing the remaining tritium to diffuse through He gas. This results in far longer tritium absorption times and can only be avoided by circulation of the gas through the absorber material. The blanketing effect may also be minimized by removal of the ^3He produced after short times, thus avoiding the generation of large ^3He concentrations in the metal and/or the gas.

The disintegration of most metal-hydride storage materials into powder with its large surface area (Sect.3.4.1) can also generate problems: 1) Metal powder is often highly pyrophoric; hence leakage, and/or exposure to oxygen or air should be avoided. 2) The very fine powder can be transported by gas streams and its spread must be avoided by filter barriers. 3) Powder needs a larger volume than solid pieces of the same mass. In the case of uranium, the powder occupies a volume at least twelve times as large as the solid. The walls of the container can be damaged after disintegration if this is not allowed for.

In the following sections, the most important property for tritium storage, namely the solubility, is presented for a few selected metals. A discussion of the isotopic effects of H, D, and T in other metals is left until Chap.5.

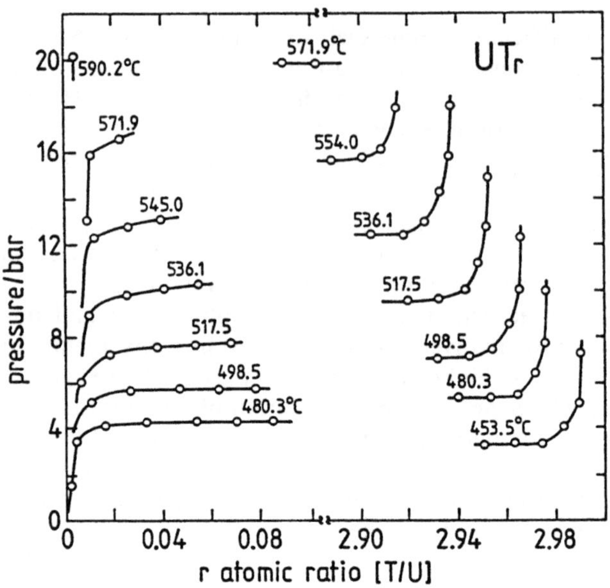

Fig.3.6. Pressure-concentration isotherms of the U-T system [3.28]

3.6.1 Uranium Tritide

Some of the properties of the U-T system were discussed in Sect.3.4.1. The pressure-concentration-temperature relationship of the U-T system [3.28] is shown in Fig.3.6. A very high tritium storage capacity of, in principle, 3 tritium atoms per uranium atom, is evident. The equilibrium pressure for $UT_{2.88}$ is described by the van't Hoff equation $\log(p/\mathrm{bar}) = -4038.2/T + 6.080$ for the temperatures between 453° and 572°C [3.28].

3.6.2 Palladium Tritide

Palladium plays a special role in the group of metal hydrides because, in contrast to most other metals, it is resistant to passivation by environmental gases due to the catalytic nature of its surface. Thus palladium is used in cases where other metal hydrides easily become poisoned. The results of such poisoning effects, due to small impurity levels in the gas, are a drastic reduction of reaction capability and absorption capacity, requiring a further heat treatment for reactivation. Thin films of palladium coated on kieselguhr yield large surface areas and high pumping speed. Pd films are especially suited to absorb tritium out of gas mixtures.

Figure 3.7 shows the pressure-concentration-temperature relationship of the Pd-T system [3.29] measured over a large pressure and temperature range. Information about isotopic effects on the solubility, diffusivity, phase boundaries and optical phonons of the Pd-X (X = X, D, T) system can be found in Chap.5.

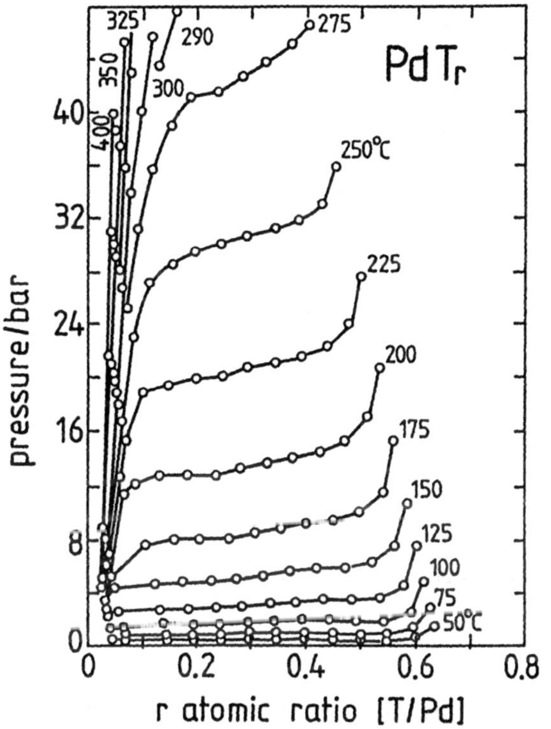

Fig.3.7. Pressure-concentration isotherms of the Pd-T system [3.29]

3.6.3 $LaNi_{5-x}Al_x$ Tritides

According to *Ortman* et al. [3.30], the tritium storage system at Savannah River Laboratory, one of the largest tritium production plants in the USA, is partly based on metal tritides, especially $LaNi_{5-x}Al_x$ tritides. The main reasons for this choice were: firstly, that at a given temperature, the absorption and desorption pressures can be tailored to the specification of the different applications by a simple variation of the Al concentration in the alloy; secondly, that the $LaNi_{5-x}Al_x$ material does not show significant disproportionation after many charging and temperature cycles in contrast to most other binary or ternary alloys. The prevention of disproportionation is very important in the case of tritium storage because otherwise tritium would become trapped in one component (for example, in the La metal) with a very low equilibrium pressure. Considering the tritium isotherms of the La-T system given in Fig.3.8, one recognises that such trapped tritium could only be recovered at a significant rate by heating the alloy to temperatures above 1000 K.

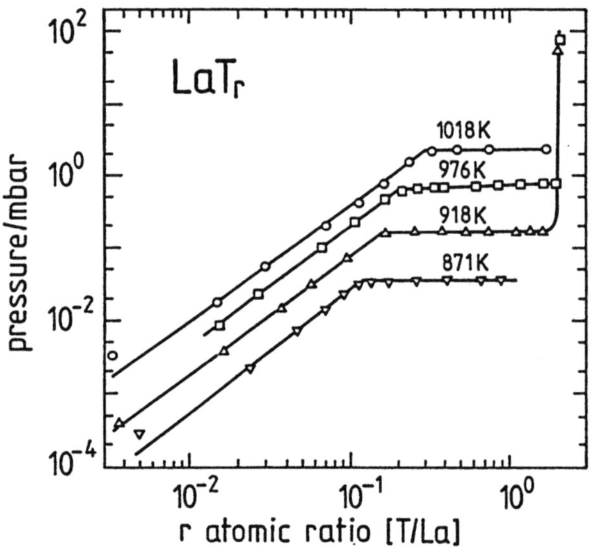

Fig.3.8. Pressure-concentration isotherms of the La-T system [3.32]

3.6.4 Vanadium Tritide

Solubility data for tritium in vanadium [3.29] exist only in the concentration range $0.8 \leq r \leq 2.0$. The isotherms show very large equilibrium pressures at relatively low temperatures (Fig.3.9). Thus VT_2 is probably the best suited material to generate very high tritium pressures. For these low temperatures (Fig.3.9) the pressure stability of the VT_2 container used as compressor can easily be calculated because the mechanical properties of most stainless steels are well known only in the temperature range below 450°C. A further advantage is that the equilibrium pressure increases only slightly over a large concentration range between r = 0.9 and 1.9. The equilibrium pressure of $VT_{1.4}$ is given by the equation [3.29] $\log(p/\text{bar}) = -2032.4/T + 7.101$ for temperatures between 126° and 210°C.

VD_2 and VDT alloys were also used by *Carstens* and *David* [3.31] to produce D_2 and DT pressures up to 900 bar for filling micro-targets for laser fusion experiments.

3.6.5 Tritides of Ti, Zr, La and the Rare Earths

The IIIB, IVB and rare-earth (RE) elements of the periodic table form very stable di- and/or tritritides. The equilibrium pressures of these metal tritides at room temperature are so low that they are hardly measurable. Thus the pressure-concentration-temperature relationship of these metal hydrides has to be determined at high temperatures. As an example Fig.3.8 [3.32] shows the isotherms of the La-T system. For concentrations lower than r = 0.1 Sieverts' law is obeyed. The horizontal plateau pressures belong

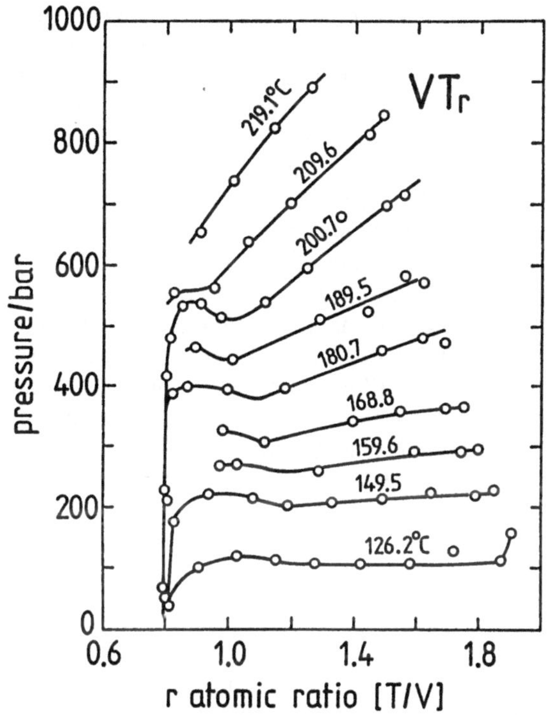

Fig.3.9. Pressure-concentration isotherms of the V-T system [3.29]

to the two-phase region of the α phase at low concentrations and the LaT_2 phase. To obtain pressures which are readily measurable in the α phase, the LaT_r sample has to be heated to above 600°C.

The metals mentioned above can clean gas mixtures of tritium to extremely low residual tritium concentrations. In addition, they are used for the final storage of tritium. To recover the tritium absorbed in these metals, the metal tritides have to be heated to high temperatures, thereby giving rise to the danger of higher tritium permeation rates at these temperatures. Therefore, these metals are mainly used for final storage as waste disposal material and rarely as reversible storage materials.

3.6.6 Stability of Binary and Ternary Metal Tritides

Having discussed a few metals used for tritium storage, some general statements concerning the long time stability of these alloys should be made. The performance of metal tritide storage containers is never as good as that of metal protide and metal deuteride storage containers. This is caused by the tritium decay product ^3He, which can generate blanketing problems in the gas phase, as mentioned above. In addition, the effects of ^3He are even more drastic in the interior of metal tritides (Chap.6). Due to their insolubility, the He atoms can produce highly pressurized bubbles. The metal atoms that previously occupied the place of the He bubbles are emitted as self-interstitial atoms (SIAs) into the metal lattice. The SIAs can cluster, produce SIA loops and an evolving dislocation network. At dislocations hydrogen atoms can become trapped and properties of metal tritide alloys may change, e.g. the solubility, diffusivity, etc., as a function of time, or increasing He concentration or dislocation density. These effects may be even stronger in ordered ternary metal tritide alloys ABT_r where the A and B atoms occupy special sites. After the emission of the atoms A and B it is highly unlikely that the atoms A and B again occupy the correct sites. The order parameter may decrease with time and, as a consequence, a new alloy may be produced with different metal tritide properties. One recognises immediately that in a binary metal tritide alloy the emission of SIAs or the punching out of dislocation loops plays a minor role because site occupancy is of no importance. Due to these possible changes of metal tritide properties, ternary metal tritide alloys should be used with caution because not enough is known about their long time behaviour.

3.7 Permeability of Tritium in Metals

As mentioned in earlier sections, the best way to contain tritium is by the principle of multiple containment using tritium compatible materials with very low permeation coefficients.

This section summarizes part of the data on permeability of hydrogen in metals. The permeation of hydrogen in metals has been studied quite often, but the agreement of the obtained values is poor. The differences are partly caused by poorly cleaned surfaces. Most of the older experimental setup for determination of permeation coefficient had no surface sensitive equipment to characterise the cleanliness of both up- and downstream surfaces. Even small amounts of oxygen at the surface can change the permeation rate. Well-grown oxide layers at the surfaces are used to reduce protium, deuterium and tritium permeation. Furthermore the permeation of hydrogen through metals is also influenced by trapping at internal defects.

In the case of metals, the fundamental permeation equation can be written as

$$\frac{dQ}{dt} = P\,\frac{A}{d}(\sqrt{p_u} - \sqrt{p_d}) \tag{3.1}$$

where dQ/dt is the number of hydrogen moles permeated per second through a sample of area A and thickness d, p_u (p_d) is the pressure of hydrogen on the upstream (downstream) side of the sample and P the permeability.

For glass the values dQ/dt is proportional to the difference of the upstream and downstream pressures because the hydrogen molecules can not be dissociated at those surfaces.

In the case of metals and glasses, the permeability can be expressed by

$$P = P_0 \exp\left(\frac{-E_p}{kT}\right), \tag{3.2a}$$

$$P = P_0 T \exp\left(\frac{-E_p}{kT}\right), \tag{3.2b}$$

respectively, where P_0 is a pre-exponential factor, E_p the activation energy for permeation and k the Boltzmann constant.

The permeability can be described as the product of solubility (given as the Sieverts' constant K^∞, Sect.5.1) and diffusivity (given as diffusion coefficient D, Sect.5.10). Therefore, in principle, the permeability can be calculated if solubility and diffusivity are known. This procedure should only be used if K^∞ and D have been measured for the same sample.

Table 3.1. Permeabilities $P = P_o\exp(-E_p/kT)$ of protium, deuterium and tritium in metal [3.33]

Permeated material	Hydrogen Isotope	P_0 [mol/(ms $Pa^{1/2}$)]	E_p [eV]	Measured temperature [K]	Reference
Al	T_2	$5.8\cdot10^{-5}$	1.28	420-520	3.35
Ag	H_2	$3.4\cdot10^{-8}$	0.633	730-980	3.36
Au	D_2	$3.1\cdot10^{-6}$	1.28	500-900	3.37
Co(α)	D_2	$3.8\cdot10^{-8}$	0.67	670-820	3.38
Co(ε)	D_2	$6.3\cdot10^{-9}$	0.59	460-670	3.38
Cu	H_2	$8.4\cdot10^{-7}$	0.80	470-710	3.39
Fe	H_2	$4.1\cdot10^{-8}$	0.36	375-850	3.40
Ge	H_2	$1.2\cdot10^{-5}$	2.07	1040-1200	3.41
Mo	H_2	$2.3\cdot10^{-7}$	0.837	500-1700	3.42
Ni	H_2	$4.0\cdot10^{-7}$	0.57	480-690	3.43
Pd	H_2	$2.2\cdot10^{-7}$	0.162	300-709	3.44
Pt	H_2	$1.2\cdot10^{-7}$	0.73	540-900	3.45
Si	H_2	$1.4\cdot10^{-5}$	2.33	1240-1485	3.41
Stainless Steel	T_2	$1.38\cdot10^{-7}$	0.684	370-700	3.46
Ta	H_2	$5.8\cdot10^{-9}$	-0.209	calculated	3.47,48
V	H_2	$4.0\cdot10^{-9}$	-0.258	calculated	3.47,48
W	H_2	$7.8\cdot10^{-7}$	1.47	1100-2400	3.49
		[mol/(msK Pa)]			
Soda-lime glass	D_2,DT	4.10^{-17}	0.48	300-700	3.50

Most permeation experiments have been performed at high temperatures to achieve easily measurable permeation rates. Extrapolation to lower temperatures must be done with care. The hereby introduced error and the disagreement of the experimental values of different groups - even for the same hydrogen isotope - are so large that the comparatively small correction expected due to isotope effects (e.g., a reduction of the permeability by the inverse square root of 3 for T compared to H according to classical theory (Sect.5.10)) can often be neglected. This is the reason why the permeability of H, D, T in metals is not discussed in Chap.5.

A careful review of many permeation experiments was given by *Steward* [3.33]. His, as best chosen values, will also be used here. Furthermore, hydrogen permeation data of "impermeable" materials are listed in the book of *Souers* [3.34]. The data used by these two researchers are presented partly in Table 3.1. The permeability of hydrogen in these metals increases in the sequence Si, Ge, W, Au, Al, Mo, Cu, Ni, Fe. The permeabilities of many stainless steel alloys are between the ones of Ni and Cu. Pd, Ta, V and other metals (rare earth metals) reacting exothermically with hydrogen, show large permeation rates, because the activation energies E_p for the permeation are small due to the high solubility.

4. Analysis of Tritium in Metals

To analyze tritium in metals, almost all conventional techniques employed for the study of the stable hydrogen isotopes in metals can be used: Secondary Ion Mass Spectroscopy (SIMS) [4.1], Auger spectroscopy [4.2], Nuclear Magnetic Resonance (NMR) [4.3], measurements of the length or lattice parameter change in comparison with pure metals [4.4], weight changes [4.5], depth profiling [4.6] using the nuclear reaction $^3H(d,n)^4He$, channeling, neutron spectroscopy [4.7], neutron radiography [4.8], electron microscopy, etc. Only certain nuclear physical methods specific to tritium will be described in this chapter. Very small amounts of tritium can be detected with techniques using proportional counters and ionisation chambers, liquid scintillation detectors and photographic emulsions [4.9, 10]. A simple method to determine large amounts of tritium in a metal bed is the measurement of the temperature increase in the bed due to the energy created by the dissolved tritium. The heat output of 1 g tritium is 0.324 W.

4.1 Outgassing and/or Combustion

Probably the highest accuracy and sensitivity of tritium detection can be obtained with liquid scintillation detectors. This means that to determine the activity of an MT_r sample, the tritium atoms have first to be bound to water. This may be achieved by burning the tritium of a MT_r sample in a Wickbold apparatus to HTO, which is in turn analysed. Another possibility is to dissolve the MT_r sample in an acid in a closed container where tritium atoms exchange with hydrogen atoms of the acid. The HT gas which also arises is oxidised to HTO in a second step, for instance using H_2O_2 and a precious metal catalyst.

To avoid the oxidation of tritium, the amount of T in a metal can be determined by heating the sample to high temperatures and then measuring either the pressure increase in a known volume with normal pressure gauges, or the outgassed activity with ionisation chambers. The purity of the tritium gas can be checked by mass spectrometry [4.11], gas chromatography [4.12] or laser Raman spectroscopy [4.13].

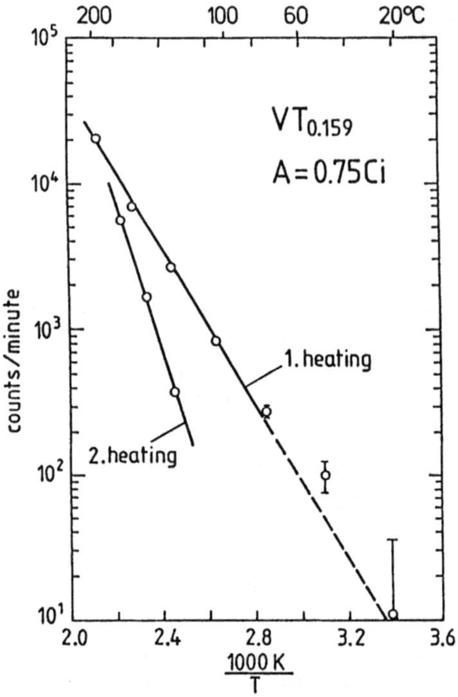

Fig.4.1. Tritium evolution of a $VT_{0.159}$ sample as a function of reciprocal temperature [4.14]. The sample was heated twice between 20 and 200°C. 10 counts/min correspond to an activity loss of 1.42 Bq/min ($3.83 \cdot 10^{-11}$ Ci/min)

4.2 Gas Release Behaviour of MT_r Samples

As briefly indicated in Sect.3.4.2, handling of MT_r samples may be required for further experiments outside the charging apparatus. Thus information on the gas release of the MT_r samples is essential for safety considerations. To date, MT_r-samples with activities of up to 37 TBq (1000Ci) have been handled in the hood of the author's tritium laboratory. This was permissible in view of the experience previously gained with MT_r samples of much lower activities. An example of a tritium gas release behaviour is shown in Fig.4.1, where the tritium permeation rate given in counts per minute as measured with a special windowless tritium monitor is plotted as a function of reciprocal temperature. A $VT_{0.159}$ single crystal (mass: 8.4mg) with an activity of 27.8 GBq (≃0.75Ci) was heated from room temperature to 200° C twice in air [4.14]. The air around the MT_r samples is pumped through a tritium monitor and afterwards through molecular sieves to absorb the water vapour in the air. The outgassing behaviour shown in Fig.4.1 can be described by an Arrhenius equation. The slopes of the straight lines yield the activation energy for the release process. The following values were obtained: 0.54 and 1.04 eV/tritium atom for the first and second run [4.14], respectively. This large increase of the activation energy probably reflects modifications of the surface layers due to the heating in air, further oxidation and/or formation of other, more stable oxides. Also, other impurities such as nitrogen, sulphur, carbon etc. may be present at the surface and influence the gas release behaviour. Given the fact that MT_r samples could not be analysed with Auger or X-ray photoemission spectroscopy techniques, a definite analysis of the surface cannot be given here. It is generally believed that oxide layers drastically hinder the recombination reaction of adsorbed (ads) tritium to tritium gas (g), $T_{ads} \rightarrow \frac{1}{2} T_2(g)$, and/or the exchange reaction $H_2O + T_{ads} \rightarrow HTO + H_{ads}$. In most of the gas-release experiments the integral tritium activity measured by the tritium monitor was in fair agreement with the total amount of tritium absorbed in the molecular sieves which was determined (after exchange with water) by liquid scintillation techniques. This result shows that the second release process described above is more probable than the evolution of tritium gas or that the desorbing T_2 molecules exchange with water vapour molecules. The tritium countrate of 10 counts per minute corresponds to an activity loss of 1.42 Bq/min ($\simeq$3.83 10^{-11}Ci/min) in Fig.4.1. Assuming time-in-dependent gas release behaviour and neglecting tritium decay the $VT_{0.159}$ sample would at room temperature lose half of the tritium in about $1.9 \cdot 10^4$ years. Assuming a diffusion-controlled release, half of the tritium would be lost in less than 20 seconds.

Tritium release from niobium is also significantly influenced by surface effects [4.15].

These studies show that most metal tritides - an exception is PdT_r - can be handled safely at room temperature in well-ventilated laboratories. The risk certainly increases with a larger surface area, temperature, tritium concentration and age of the MT_r samples. On the other hand, the oxide layers on the surfaces cannot be considered absolute permeation barriers for tritium. This result can probably be extended to MH_r and MD_r samples.

4.3 Tritium Induced X Rays

According to (1.1) the maximum energy of the β^- electron generated by the tritium decay is 18.582 keV. If tritium atoms dissolved in a metal decay, the created β^- electrons can eject electrons from the deep shells with binding energies below 18.582 keV of the surrounding host lattice atoms. Alternatively, they can produce secondary electrons. The holes created are filled with electrons from outer shells with lower binding energies. Characteristic X rays may be emitted with an energy equivalent to the difference of the binding energies. A competing process is the production of an Auger electron. In addition, the β^- electrons may be accelerated near the nucleus of the lattice atoms and continuous Bremsstrahlung radiation may be emitted. The superposition of both X-ray contributions is presented in Figs.4.2,3 for a $VT_{0.42}$ [4.16] and a $TaT_{0.012}$ [4.17] sample,

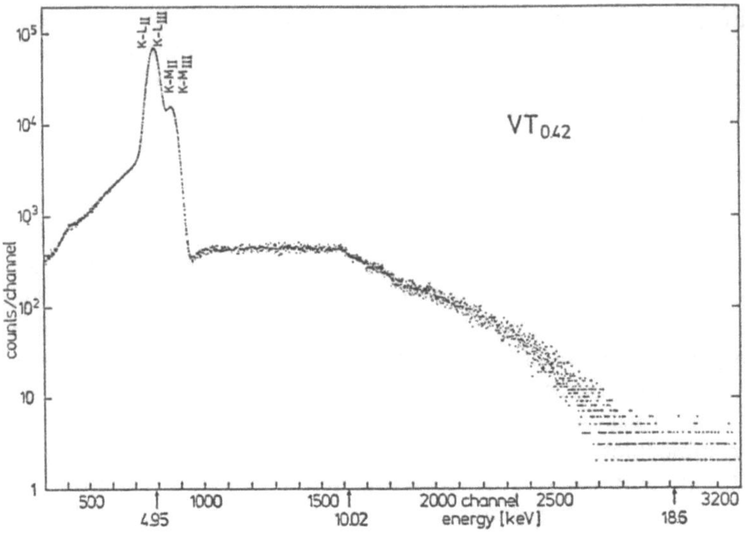

Fig.4.2. Tritium induced X-Ray spectrum of $VT_{0.42}$ [4.16] measured with an intrinsic Ge detector

respectively. The X rays were analyzed with an intrinsic Ge detector cooled to liquid-nitrogen temperature, and the usual electronics consisting of a preamplifier, main amplifier, Analog-to-Digital Converter (ADC) and MultiChannel Analyser (MCA). The counts per channel are plotted on a logarithmic scale as a function of the channel number. A few X-ray energies are indicated.

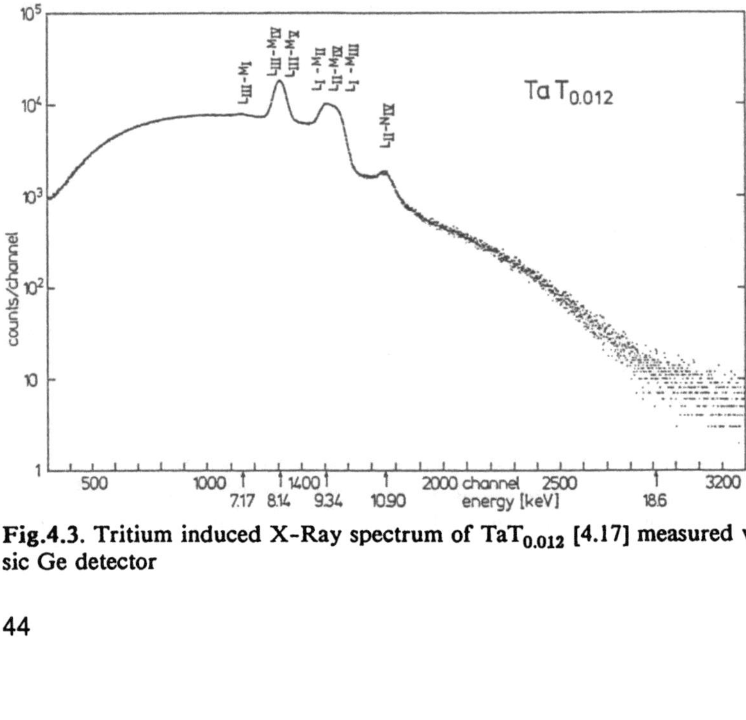

Fig.4.3. Tritium induced X-Ray spectrum of $TaT_{0.012}$ [4.17] measured with an intrinsic Ge detector

The peaks can be attributed to characteristic X-ray lines whereas the broad, flat spectrum is due to Bremsstrahlung radiation. A very strong decrease of the X-ray intensity is observed in the energy range just below 18.6 keV, which reflects the decreasing probability for production of β^- electrons with very high energy. The efficiency of the Ge detector used is constant in the measured energy range (Figs.4.2 and 3). A very accurate determination of the intensity decrease near 18.6 keV by means of a high precision X-ray diffractometer with an energy resolution of $2.5 \cdot 10^{-4}$ may offer the possibility to determine or to give an upper value to the mass of the antineutrino - a still open question in particle physics studied by many different groups [4.18]. The very few counts above 18.6 keV represent the background of the cosmic radiation. The double peak at an energy of 4.95 keV belongs to the characteristic K_{α_1,α_2} and K_{β_1,β_3} X rays of vanadium in Fig.4.2, whereas the peaks of the $TaT_{0.012}$ X-ray spectrum can be attributed to certain L X rays indicated in Fig.4.3. No K X rays are excited in the case of the $TaT_{0.012}$ sample because the binding energy of the 1s electrons of Ta is much higher than the energy produced by the tritium decay.

Similar spectra can be obtained for all MT_r samples. X-ray spectra of tritided erbium films on molybdenum substrates are presented in [4.19]. These spectra demonstrate that if tritium is absorbed within the X ray escape length in solids, its presence can easily be detected with standard X-ray monitors used in radioactive control areas. This is in contrast to the difficult detection of tritium in air. In tritium decay the β^- electrons cannot penetrate even very thin foils, thus windowless tritium monitors are required. The disappearance of the X-ray intensity above 18.6 keV is a good indication for the presence of tritium in solids.

4.4 Tritium Imaging

Tritium imaging is based on the detection of the secondary electrons produced by the tritium β^- decay in solids (Sect.4.3). The secondary electrons produced in MT_r samples within a surface layer of about 200 nm thickness may reach the surface and can be focussed in vacuum by electrostatic lenses onto a microchannel plate image intensifier. A visual display of the intensity of the secondary electrons, or of the tritium concentration in the surface layer, is then obtained with a phosphor-coated fibre optic screen. The ultimate lateral resolution of this technique, developed by *Malinowski* [4.20-22], is about 200 nm due to the projected electron range of the averaged β^--decay energy of 5.7 keV. The resolution limit is thus about the same as in tritium autoradiography [4.23,24]

The main advantages of the tritium imaging technique are [4.20-22]: 1) its sensitivity, which is about the same as in SIMS; 2) simplicity of the setup; 3) it is nondestructive: imaging requires neither sputtering of surface layers as in SIMS nor ion implantation as in Nuclear Reaction Analysis profiling (NRA) [4.6] which can be used to probe tritium in a depth up to 20 μm below the surface by means of T(p, n) reactions; 4) two-dimensional pictures are obtained. The main disadvantages are: 1) difficulties in determining tritium concentrations from the above-mentioned imaging technique; 2) the dependence of secondary electron emission on the surface conditions; and 3) the samples can only be studied under vacuum. Very recently a new imaging system consisting of a scintillation screen, two tapered fibres, an image intensifier and a Polaroid film for image recording has been described [4.25], which needs no vacuum.

With regard to the detection of tritium, the new imaging techniques are certainly an interesting addition to the existing techniques of SIMS and NRA.

4.5 Tritium Autoradiography

Tritium near the surface of solids can be detected with photographic emulsions [4.26] exposed to the primary, backscattered and secondary electrons, Auger electrons and X rays induced by the β^- decay. Rather sophisticated autoradiographic techniques [4.23, 24, 27-29] have been developed to obtain information from the exposed areas of the photographic emulsion about the location, the distribution and the amount of tritium at the surface. The resolution of tritium autoradiography is better than for other β^- emitters due to the very low β^--decay energy of tritium and the corresponding small range. A measurable response in nuclear emulsions above background is obtained by exposures of 10^7 to 10^8 electrons/cm^2. For a recent review article about tritium autoradiography, see [4.10] by *Caskey*.

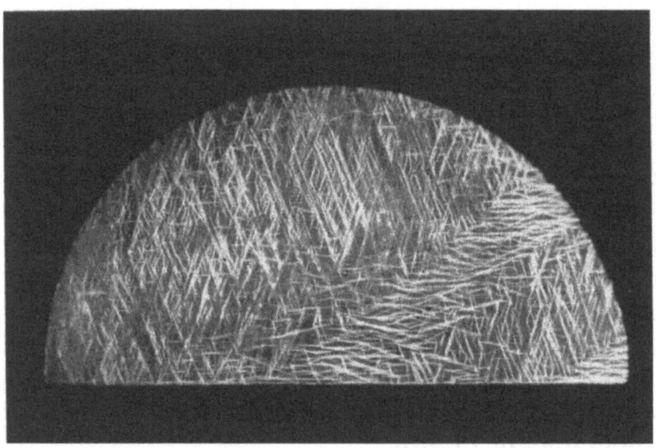

Fig.4.4. Tritium induced autoradiographic picture of a polished $VT_{0.10}$ sample (diameter: 1.2 cm) [4.17]. The bright needles are the images of the β phase domains with a tritium concentration of 50% and the grey background belongs to the α phase with about 4% tritium

As a simple example of contact autoradiography, a micrograph of a $VT_{0.10}$ sample taken at room temperature [4.17] is shown in Fig.4.4. A thin, X-ray sensitive photographic emulsion placed on a glass substrate (Ilford L4) was exposed to the polished surface of a $VT_{0.10}$ sample. The sample contains two phases with different tritium concentration at room temperature: an α phase with about 4% and a β phase with about 50% (Sect.5.5.2). The bright needles in Fig.4.4 are the images of the β phase domains arising from the about 10 times more intensive exposure of the $VT_{0.50}$ regions than in the $VT_{0.04}$ areas. The diameter of the polished sheet was 1.2 cm. The elastic distortion energy is minimized by the generation of the β phase domains seen as sharp needles in Fig.4.4.

Tritium autoradiography [4.10] was applied successfully to the study of diffusion, solubility, distribution and trapping of tritium in metals and alloys. With electron microautoradiography, a magnification of 10^4 of microscopic features can be obtained [4.10]. In Sect.5.5.2 a new application of autoradiography to determine MT_r phase diagrams will be presented.

5. Properties of Protium, Deuterium and Tritium in Selected Metals

In this chapter the isotopic dependence of physical properties obtained in recent years will be discussed for different M-X (X = H, D, T) systems.

5.1 Solubility of Protium, Deuterium and Tritium in Palladium

Probably the most frequently studied metal hydrides and deuterides are those of Pd. These systems have been reviewed a few times recently [5.1-3]. Although the properties of the Pd-H and Pd-D systems have been studied for more than 120 and 50 years, respectively, reliable information on the Pd-T system, such as solubility, superconductivity, phase boundaries, optical phonons, etc. has been obtained only in recent years.

5.1.1 Solubility of Protium, Deuterium and Tritium in the α Phase of the Pd-X (X=H, D, T) System

One of the frequently studied properties is the solubility behaviour of H and D in Pd [5.4-18]. In contrast, little and conflicting information exists for the solubility of tritium in Pd in very limited temperature (T), concentration (r) and pressure (p) ranges (for Pd black: $200 \le T \le 400$ [°C], $7 \le p \le 20$ [mbar], $7.3 \cdot 10^{-4} \le r \le 0.0047$ [5.19]; for bulk Pd plates of 120 μm thickness: $25 \le T \le 70$ [°C], $1.3 \le p \le 50$ [mbar], $0.0024 \le r \le 0.014$ [5.20]; for Pd foils of 10 μm thickness: $35 \le T \le 125$ [°C], $100 \le p \le 2000$ [mbar], $0.02 \le r \le 0.65$ [5.21,22] and for Pd wire with 1.2 mm diameter: $60 \le T \le 400$ [°C], $0.016 \le p \le 1.3$ [bar], $0.0015 \le r \le 0.02$ [5.23,24]. Only one solubility experiment of the two-phase region (Fig.3.7) was performed in a large pressure range extending to 48 bar [5.25]. No description of how the data were determined (whether in absorption or desorption) was given.

The scarcity of information on tritium solubility in Pd is mainly due to the risks in handling tritium. The partly contradictory results are probably caused by the lack of accurate data on purity of the T_2 gas and by the assumptions made for the extrapolation to 100% pure T_2 gas. This lack of information may be one reason why the solubility data for T in Pd measured by *Schmidt* and *Sicking* [5.20] and reproduced in a recent review article by *Sicking* [5.26] together with the solubility data of H and D determined by *Wicke* and *Nernst* [5.7] agree almost with the D solubility data. This is in contrast to recent solubility measurements [5.23, 24] performed for H, D and T in Pd using the same sample and identical experimental conditions.

Small amounts of the stable isotopes can influence the true tritium equilibrium pressures, as shown by desorption isotherms for the α phase of the Pd-X (X = H, D, T) system [5.23] in Fig.5.1. The squares denote the solubility data obtained for a tritium gas mixture with the composition of 93.4 at.% T, 1.9 at.% D and 4.7 at.% H. The circles with the labels H and D were determined with H_2 and D_2 gases of 99.9999 vol.% and 99.7 vol.% purity, respectively. The experimental data measured with the mixture of

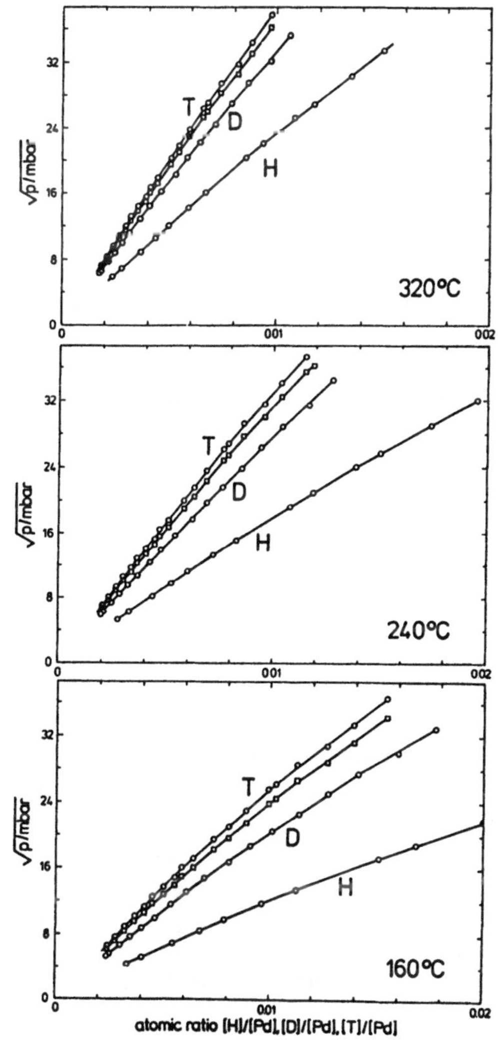

Fig.5.1. Desorption isotherms for the pure hydrogen isotopes (o) protium (H), deuterium (D), tritium (T) and the isotope mixture (93.4 at.% T, 1.9 at.% D, 4.7 at% H) (□) in the α phase of the Pd-X (X = H, D, T) systems [5.23]

isotopes given above were corrected to 100% T_2 gas by means of a program [5.27] which 1) considers that the measured pressure is the sum of the partial pressures of H_2, HD, HT, D_2, DT, T_2 molecules, 2) uses the experimentally determined relationship between p, T and r for pure H_2 and D_2, 3) assumes that the partial pressures of the six different molecules are in thermodynamic equilibrium, and 4) takes Raoult's law into account to calculate the absorbed amount of each hydrogen isotope. As expected, the correction shifts the experimental data (as given by the squares) to higher pressure values. From the data plotted in Fig.5.1 for H, D and T, the continuing decrease of the solubility of tritium in Pd in comparison with D and H is immediately noticeable. The following inequalities [5.23]

$$r_H > r_D > r_T \tag{5.1}$$

are fulfilled for constant temperature and pressure. For constant concentrations and temperature one has

$$p_{H_2} < p_{D_2} < p_{T_2} . \tag{5.2}$$

In addition, the three curves in Fig.5.1 demonstrate that the solubility of tritium decreases with increasing temperature, which means that tritium is absorbed in Pd exothermically like protium and deuterium.

Equilibrium conditions between the absorbed hydrogen concentrations r_X and applied X_2 (X = H, D, T) pressures for the desorption process $PdX_r \rightarrow Pd + \frac{1}{2} rX_2$ (which were found to be equal in the α phase to the absorption pressures within the experimental accuracy) are calculated from the equality of the chemical potentials of hydrogen in the gaseous state

$$\frac{1}{2}\mu_{X_2}(p,T) = \frac{1}{2}\left[H^0_{X_2} - TS^0_{X_2} + RT\ln(p/p_0)\right] \tag{5.3}$$

and in the solid solution

$$\mu_X(r,T) = \bar{H}^\infty_X - T\bar{S}^{\infty,nc}_X + RT\ln\left(\frac{r}{N-r}\right) + \Delta\mu_X , \tag{5.4}$$

where $H^0_{X_2}$ and $S^0_{X_2}$ are the standard molar enthalpy and entropy of the gas phase at temperature T and pressure p_0 = 1.013 bar; $\bar{H}^\infty_X$ and $\bar{S}^{\infty,nc}_X$ are the partial molar enthalpy and nonconfigurational (nc) entropy of one mole X (X = H, D, T) atoms at infinite (∞) dilution in Pd. N is the maximum number of interstitial sites over which the X atoms are statistically distributed per metal atom and $\Delta\mu_X$ is an excess chemical potential term describing deviations from the ideal absorption behaviour. Setting N = 1 for PdX_r, using (5.3, 4) and assuming that $\Delta\mu_X$ is proportional to r, the following result is obtained

$$\ln\left[\frac{p}{p_0}\left(\frac{1-r}{r}\right)^2\right] = 2\cdot\ln K^{\infty} + \frac{2}{RT}\Delta\mu^* r \qquad \text{with} \tag{5.5}$$

$$2\cdot\ln K_X^{\infty} = \frac{\Delta\bar{H}_X^{\infty}}{RT} - \frac{\Delta\bar{S}_X^{\infty}}{R}, \tag{5.6}$$

$$\Delta\bar{H}_X^{\infty} = 2\bar{H}_X^{\infty} - H_{X_2}^0, \tag{5.7}$$

$$\Delta\bar{S}_X^{\infty} = 2\bar{S}_X^{\infty,nc} - S_{X_2}^0. \tag{5.8}$$

Here K^{∞} is the Sieverts' constant, and R is the gas constant. $\Delta\bar{H}_X^{\infty}$ and $\Delta\bar{S}_X^{\infty}$ are the partial enthalpy and entropy of solution of hydrogen in a metal per mole X_2 molecules at infinite dilution.

As an example, the solubility of D in Pd obtained during heating of a PdD_r sample between 60 and 400°C in 20° steps is shown in Fig.5.2. The values on the left side of (5.5) are plotted as a function of concentration r. It is apparent that the experimental data in Fig.5.2 are well described by straight lines given by (5.5), which intersect the y axis at $2\cdot\ln K^{\infty}$. These values are plotted in Fig.5.3 as open squares [5.23] with the data of *Powell* [5.28] shown as open circles and also the results of other groups [5.7, 10] for the three pure hydrogen isotopes. Excellent agreement for H and D is obtained in the independent measurements of *Lässer* and *Powell* [5.28]. The

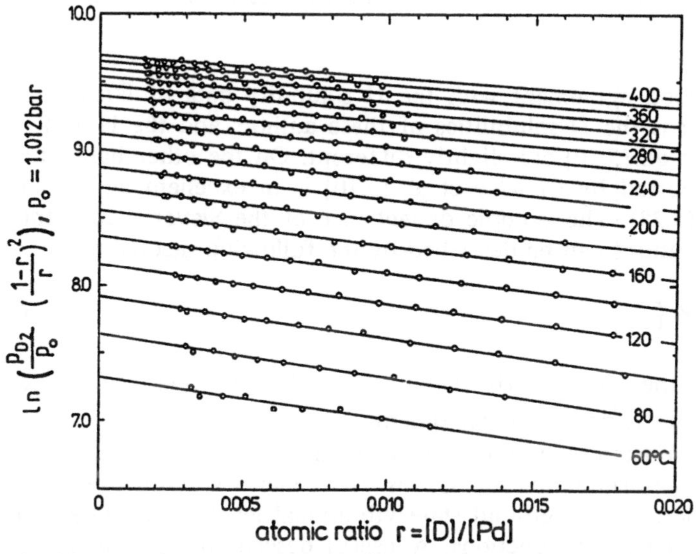

Fig.5.2. Plot of $\ln[p_{D_2}/p_0\cdot((1-r)/r)^2]$ [left-hand side of (5.5)] as a function of concentration r in the α phase of the Pd-D system

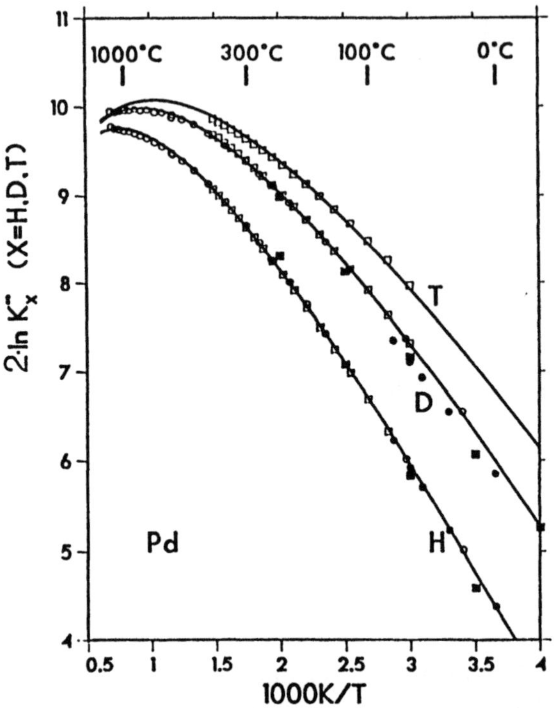

Fig.5.3. Sieverts' constant K_X^{∞} for the hydrogen isotopes protium (H), deuterium (D) and tritium (T) in Pd determined independently by *Lässer* (□) and *Powell* (o) [5.28]. In addition, data of *Wicke* and *Nernst* (●) [5.7] and *Clewley* et al. (■) [5.10] are plotted

data are also in good agreement with the H solubility data of *Wicke* and *Nernst* [5.7], whereas the D solubilities of these authors and the H and D data of *Clewley* et al. [5.10] are slightly different. On the other hand, the isotopic effect reported by *Clewley* et al. [5.10] is in excellent agreement with the data of [5.28]. The isotopic dependence of the Sieverts' constants at identical temperatures can be described by the following inequality:

$$K_H^{\infty} < K_D^{\infty} < K_T^{\infty} . \tag{5.9}$$

The curvature of the plots clearly demonstrates that the quantities $\Delta\bar{H}_X^{\infty}$ and $\Delta\bar{S}_X^{\infty}$ of (5.6) are temperature dependent.

The Sieverts' constant can be described by analytical expressions [5.28] which allow the determination of Gibbs free energies, enthalpies and entropies of solution, of the ground state vibrational levels of H, D and T in Pd and of the partition functions. A useful description of the partial Gibbs free energy of one mole of hydrogen atoms in a metal is the following equation

$$\bar{G}_X^{\infty}/(RT) = -\ln\Xi_X = -\ln(N\zeta\xi) - (E-E_0)/T \tag{5.10}$$

given by *Powell* [5.29]. Ξ is the total partition function of hydrogen dissolved in a metal, ξ is the partition function based on a well-defined vibrational model referenced to its lowest energy state (ground state), and ζ is the partition function of all other energy states of the system not included in ξ. The term ζ is approximated by

$$\zeta = (1 + Ae^{-B/T}) \tag{5.11}$$

where A represents the degeneracy of energy levels at energy B. This expression may describe anharmonic vibrational contributions, other vibrational manifolds and electronic states. The term $(E-E_0)$ in (5.10) is the ground state energy of hydrogen. E is the energy of the vibrational potential minimum measured with respect to atomic hydrogen, and E_0 is the zero point energy of the oscillator.

Probably the best description of the standard Gibbs free energy of one mole of hydrogen gas in the temperature range above 200 K is given by [5.29]

$$G^0_{X_2}/(RT) = -\ln\Xi_{X_2} = -\ln\left(\frac{LT^{7/2}}{1-e^{-J/T}}\right) - \frac{M}{T} \quad \text{with} \tag{5.12}$$

$$M = D^0_0 + B_0/3\ . \tag{5.13}$$

L, J, D^0_0 and B_0 are constants describing the hydrogen molecules. J is the Einstein temperature of the gas molecule, D^0_0 its dissociation energy (e.g., the ground state energy of the gas molecule relative to the atoms at rest) and B_0 the rotational constant of the X_2 molecules. Most of the translational rotational partition function is represented by $LT^{7/2}$. The values of the constants L, J, M, D^0_0 and B_0 are given in Table 5.1 [5.28]. A few of the energy values shown in Table 5.1 and in the following formula are

Table 5.1. List of the parameters L, J, M, D^0_0 and B_0 [5.28] to calculate the standard Gibbs free energy of one mole of diatomic hydrogen gas with (5.12, 13)

Molecule	$L[K^{-7/2}]$	J [K]	M [K]	D^0_0 [K]	B_0 [K]
HH	$4.293\cdot10^{-4}$	5986	51994.9	51966.5	85.348
HD	$2.093\cdot10^{-3}$	5225	52402.2	52380.8	64.269
HT	$3.614\cdot10^{-3}$	4940	52552.1	52532.9	57.187
DD	$2.406\cdot10^{-3}$	4307	52888.2	52873.9	43.027
DT	$8.041\cdot10^{-3}$	3948	53076.1	53064.1	35.927
TT	$6.582\cdot10^{-3}$	3548	53285.6	53276.0	28.819

given in Kelvin, they can easily be converted to energy scales by multiplication with the Boltzmann constant.

For a harmonic model, the three-dimensional, isotropic, vibrational partition function and the zero point energy are given by

$$\xi = \left(\sum_{n=0}^{\infty} e^{-nC/T}\right)^3 = (1 - e^{-C/T})^{-3} \quad \text{and} \tag{5.14}$$

$$E_0 = \frac{3}{2} C , \tag{5.15}$$

respectively, where C is the Einstein temperature of the oscillator.

Combining equations (5.6,10-15) gives the following analytical expression for the Sieverts' constant:

$$2 \cdot \ln K^{\infty} = \frac{\ln(LT^{7/2})}{1 - e^{-J/T}} + \frac{M}{T} - 2 \cdot \ln \frac{N(1 + Ae^{-B/T})}{(1 - e^{-C/T})^3} - 2 \frac{E - 3C/2}{T} . \tag{5.16}$$

With (5.16) the quantities $\Delta\bar{H}^{\infty}$ and $\Delta\bar{S}^{\infty}$ can be calculated in an analytical form using the equations

$$\Delta\bar{H}_X^{\infty} = 2R \left.\frac{\partial \ln K_X^{\infty}}{\partial(1/T)}\right|_{p,x} \quad \text{and} \tag{5.17}$$

$$\Delta\bar{S}_X^{\infty} = \Delta\bar{H}_X^{\infty}/T - 2R \cdot \ln K_X^{\infty} . \tag{5.18}$$

The unknown parameters in (5.16) were determined by a non-linear least squares procedure. To reduce the number of parameters, the following approximations were made: N was set to unity, the C values were chosen close to those determined by inelastic neutron spectroscopy. Thus, only the A, B and E values were determined by the nonlinear least squares program. Results of such a fit based on the relations $C_D = C_H/\sqrt{2}$ and $C_T = C_H/\sqrt{3}$ are given in Table 5.2.

The solid lines plotted in Fig.5.3 through the experimental points shown as open circles and squares are obtained with the parameter sets 1, 2 and 4 of Table 5.2 which will be used in the following.

Rush et al. [5.30] observed higher vibrational excitations of H and D atoms dissolved in the α-phase of Pd and described their data by adding an anharmonic term to the harmonic potential. They concluded that the potential well is isotope independent and that the vibrational energy levels E^H_{nml} of the protium atoms can be described by the equation

$$E^H_{nml} = \omega_H \left(n+m+l+\frac{3}{2}\right) + \beta_H \left(n^2+m^2+l^2+n+m+l+\frac{3}{2}\right) , \tag{5.19}$$

Table 5.2. List of the parameters A, B, C, E, N used to calculate Sieverts' constant with (5.16) for Pd-X (X = H,D,T). Dataset 4 was determined for T by scaling the parameters for H and D inversely as the square root of the isotopic mass [5.28]

No	X	N	A	B [K]	C [K]	E [K]	(E-1.5C) [K]
1	H	1	1.981	768.0	800.0	28145.0	26945.0
2	D	1	1.933	664.0	565.7	28175.4	27326.9
3	T	1	2.280	677.0	461.9	28182.3	27489.5
4	T	1	1.912	617.9	461.9	28188.9	27496.0

where n, m, l denote the oscillator excitations in the main cubic directions of the Pd lattice, $\omega_H = 580.2$ K, $\beta_H = 110.2$ K, $\omega_D = \omega_H/\sqrt{2}$ and $\beta_D = \beta_H/2$. With this expression, the vibrational partition function

$$\xi = \left(\sum_{n=0}^{\infty} \exp\{-n[\omega + (n+1)\beta]/T\}\right)^3 , \qquad (5.20)$$

and the zero point energy

$$E_0 = \frac{3}{2}(\omega + \beta), \qquad (5.21)$$

can be calculated for all three isotopes, using $\omega_T = \omega_H/\sqrt{3}$ and $\beta_T = \beta_H/3$. Inserting (5.20,21) instead of the purely harmonic model into (5.16) and fitting the new expression to the experimental data yields values for A and B different from those of the harmonic model. Values for E and $(E-E_0)$, however, are in good agreement due to the fact that at low temperatures the fit of (5.16) must be achieved by variations of the $(E-E_0)$ values because the third term on the right-hand side of (5.16) is more or less constant at low temperatures.

Averaging all the values for $(E-E_0)$ obtained by different fitting procedures [5.28] yields data for the ground states of the hydrogen isotopes in Pd at infinite dilution, which are located for H, D and T at 2322.6±1.7, 2355.2±1.1, and 2369.2±2.5 meV below the dissociation limit for half a mole of H_2 gas.

With (5.17,18) and the parameters of Table 5.2 the quantities $\Delta\bar{H}_X^{\infty}$ and $\Delta\bar{S}_X^{\infty}$ become analytical functions of the temperature. In Figs.5.4,5 $\Delta\bar{H}_X^{\infty}$ and $\Delta\bar{S}_X^{\infty}$ are presented for the isotopes X = H,D,T. In addition, values by *Boureau* and coworkers [5.12-14] determined calorimetrically are included and are in good agreement with the solid lines considering the different methods. The value $\Delta\bar{H}_D^{\infty}$ (□) given in [5.12] was shifted according to the improvements made in [5.13] and also assuming that the difference $\Delta\bar{H}_H^{\infty} - \Delta\bar{H}_D^{\infty}$ is not affected. All curves of $\Delta\bar{H}_X^{\infty}$ and $\Delta\bar{S}_X^{\infty}$ (X = H, D,

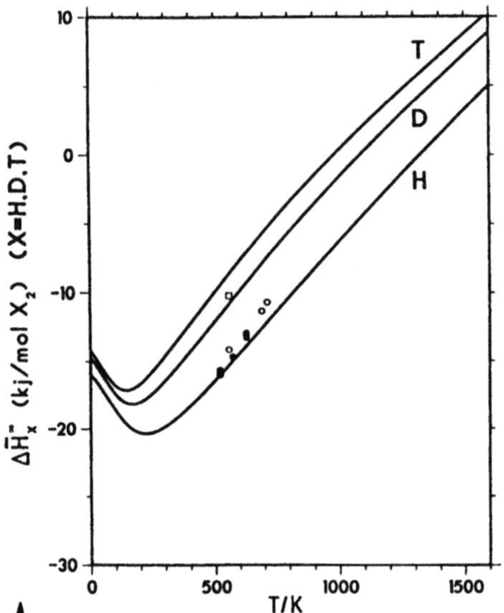

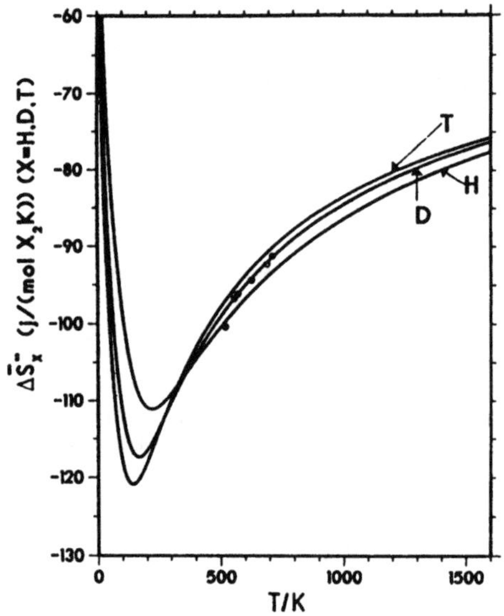

▲

Fig.5.4. The partial enthalpy $\Delta\bar{H}_X^{\infty}$ per mole X_2 for the dissolution of X (X = H, D,T) in Pd at infinite dilution [5.28]. (o: calorimetric data of H from [5.13]; □: calorimetric data of D from [5.12] with correction mentioned in the text; ●: calorimetric data of H from [5.14])

Fig.5.5. The partial entropy $\Delta\bar{S}_X^{\infty}$ per mole X_2 for the dissolution of X (X = H,D,T) in Pd at infinite dilution [5.28]. (o: calorimetric data of H from [5.13]; □: calorimetric data of D from [5.12] with correction mentioned in the text; ●: calorimetric data of H [5.14])

T) have minima corresponding to the temperature range where the thermal population of vibrational states of the H, D and T atoms in Pd becomes important. The main isotopic dependence of $\Delta\bar{H}_X^{\infty}$ and the solubility curves can be understood qualitatively by comparing the zero-point energies of the vibrational modes of the molecules and of the interstitial atoms, as plotted in Fig.5.6 on the right-hand side. It is seen that more energy is gained by the absorption of a H_2 molecule than by that of a T_2 molecule. Due to the steep slope of the potential well for the X_2 (X = H,D,T) molecules the energy differences between the vibrational levels in the gas phase are more than twice as large as for the potential well on the right side of Fig.5.6 describing the potential well of H, D and T atoms in α-PdH_r. Tritium in contact with Pd is therefore biased to remain in the gas phase compared with protium. H atoms are absorbed more readily than T atoms. A comparison of the two potential wells on the left side of Fig.5.6 shows that the opposite isotopic effect (i.e., that T atoms are absorbed more readily than D and H atoms) is expected for metals where the energy difference between the vibrational levels of the H, D, T atoms is larger than half of the difference between the corresponding H_2, D_2 and T_2

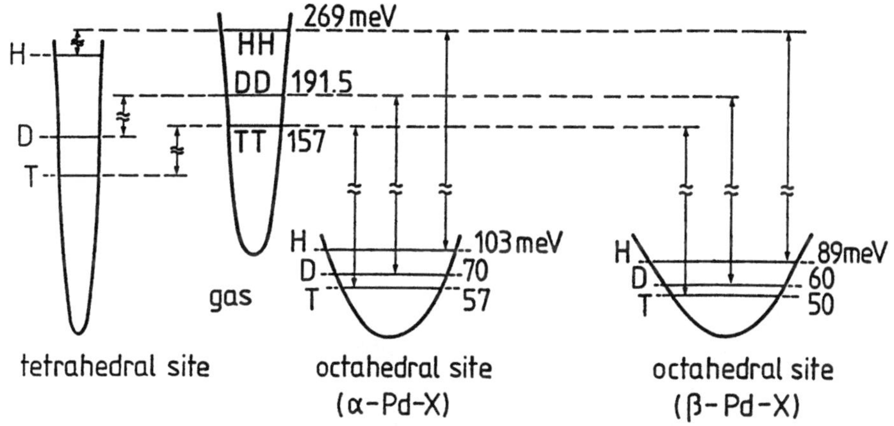

Fig.5.6. Potential wells and zero-point energies of the molecules H_2, D_2, T_2 in the gas phase and of the atoms H, D, T at octahedral (right-hand side) and tetrahedral (left-hand side) sites in metals. The zero-point energies of the atoms are scaled up per atom for the threefold degeneracy and must be counted twice for comparison with the molecular energies

molecules. With increasing temperature, first only the rotational levels in the gas phase will be populated. This results in more exothermic solution enthalpies until the temperature is reached where the population of the next higher vibrational level in Pd is significant. The change of the $\Delta\bar{H}_X^\infty$ (X = H, D, T) values at temperatures above 250 K is attributed mainly to the vibrational behaviour of the H, D and T atoms in the metal, since the first excited vibrational levels in the gas phase are mainly populated at very high temperatures (Table 5.1).

According to (5.10) the total partition function Ξ of hydrogen in metals can be separated into two terms. The first term describes the contribution of the sum of the thermally populated energy states, the second term the contribution of the ground state. The temperature dependence of the $\ln(\varsigma\xi)$ terms using the parameters of Table 5.2 is plotted in Fig.5.7 for each of the three isotopes. The contributions increase in the sequence H, D, T given the larger numbers of populated states for T than for H at the same temperature.

A rather simple description of the isotope effect of H and D in metals [5.31-33] assumes that the only isotope-dependent contribution stems from the localized vibrational modes. In Fig.5.8 the results of this model are plotted as dash-dotted lines assuming a three-dimensional harmonic oscillator. The following vibrational energies were used: $\hbar\omega_H$ = 69.0 meV [5.30], $\hbar\omega_D = \hbar\omega_H/\sqrt{2}$, $\hbar\omega_T = \hbar\omega_H/\sqrt{3}$. The logarithm of the ratio of the Sieverts' constants is shown as a function of reciprocal temperature. The experimental values (circles) and estimated error bars are plotted. Only qualitative agreement is obtained. Using the vibrational energy values determined by *Rush* et al. [5.30], see (5.19-21), only slightly better agreement is obtained,

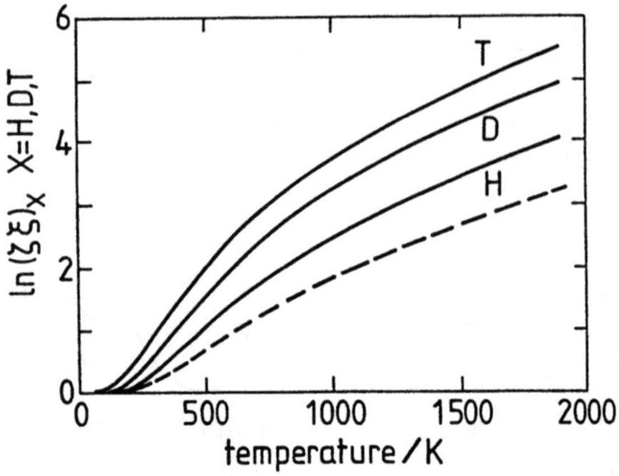

Fig.5.7. Plot of ln(ζξ), see (5.10), for H, D, T in Pd as a function of temperature using the parameters given in Table 5.2. The dashed line is obtained for H using ζ=1 and from (5.14) [5.28]

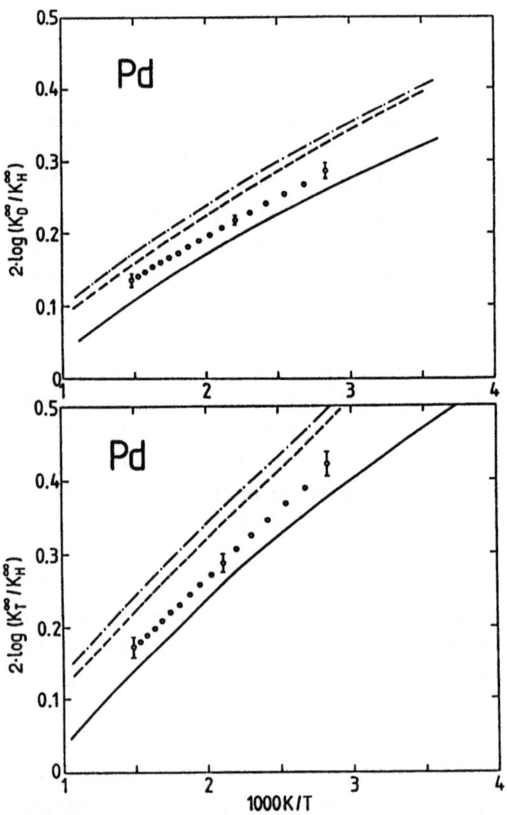

Fig.5.8. Ratio of the Sieverts' constants (K_D^∞/K_H^∞) and (K_T^∞/K_H^∞) in Pd (o: experimental values [5.23]) compared with different models (for details see text)

as demonstrated by the dashed lines. *Oates* and *Flanagan* [5.11] obtained good agreement assuming that $\hbar\omega_D = \hbar\omega_H/1.53$ and allowing for localized-mobile transitions. The recently determined vibrational energies ($\hbar\omega_H$ = 69.0meV and $\hbar\omega_D$ = 46.5meV) [5.30] do not yield the ratio used by *Oates* and *Flanagan* [5.11]. A "harmonic" model using the vibrational energies $\hbar\omega_H$ = 69.0 meV, $\hbar\omega_D$ = 46.5 meV and $\hbar\omega_T$ = 37 meV (Sect.5.8) is represented by the solid curves in Fig.5.8. Again, no satisfactory agreement is obtained. The following reasons may be given: 1) it is incorrect to consider only the vibrational contributions as isotope dependent. In Sect.5.9 it will be seen that indeed a very small isotope dependence of the electronic contributions was determined in PdX_r (X = H, D, T). 2) In Sect.5.8 the strong dispersion of the optical modes is mentioned. In the harmonic models above, the dispersion is neglected. 3) The extrapolation to higher vibrational energy values using the data of *Rush* et al. [5.30] may lead to inaccurate results. It would be very helpful to determine experimentally higher excitations than in the work of *Rush* et al. [5.30]. A technique for measuring very high vibrational levels was recently published by *Hempelmann* et al. [5.34].

So far, only the solubility of H, D and T in Pd at very low concentrations has been discussed. In the following, the solubility data of T in Pd for concentrations $r \geq 0.02$ will be compared with that of D and H.

5.1.2 Solubility of Protium, Deuterium and Tritium in the α, $\alpha+\beta$ and β Phases of the Pd–X (X=H, D, T) System

Representative isotherms [5.35] of the three hydrogen isotopes in Pd are plotted in Fig.5.9. The squares give the desorption values of an impure tritium gas mixture containing 95.2 at.% T, 1.1 at.% D and 3.7 at.% H as analyzed by mass spectrometry. The isotherms of tritium of 100% purity were obtained by extrapolation with a program mentioned above. In earlier publications [5.21, 22] no purity corrections were made. The shape of the Pd-T isotherms is similar to that of the Pd-D and Pd-H systems [5.7-9]. The Pd-T isotherm can be separated into three parts.

The horizontal lines correspond to the $\alpha+\beta$ phase region and are characteristic of the coexistence of two solid phases with the gas phase (Gibbs' phase rule). On the left-hand side of the miscibility gap the α phase exists, and here, to a first approximation, Sieverts' law is fulfilled, see (5.5). In the pure β phase (right-hand side of Fig.5.9 for $r \geq 0.57$) further uptake of hydrogen is only achieved by relatively large increases of the gas pressure. Due to the limited amount of tritium only pressures up to 2 bars could be achieved. In Fig.5.9 the same isotopic dependence of the desorption equilibrium pressure in the two-phase region $\alpha+\beta$ and the pure β phase is observed as expressed for the α phase by inequality (5.2). Measured isotherms obtained by extrapolation to 100% purity are shown in Fig.5.10 for PdT_r. The desorption plateau pressures of T, D and H are shown in Fig.5.11 as a

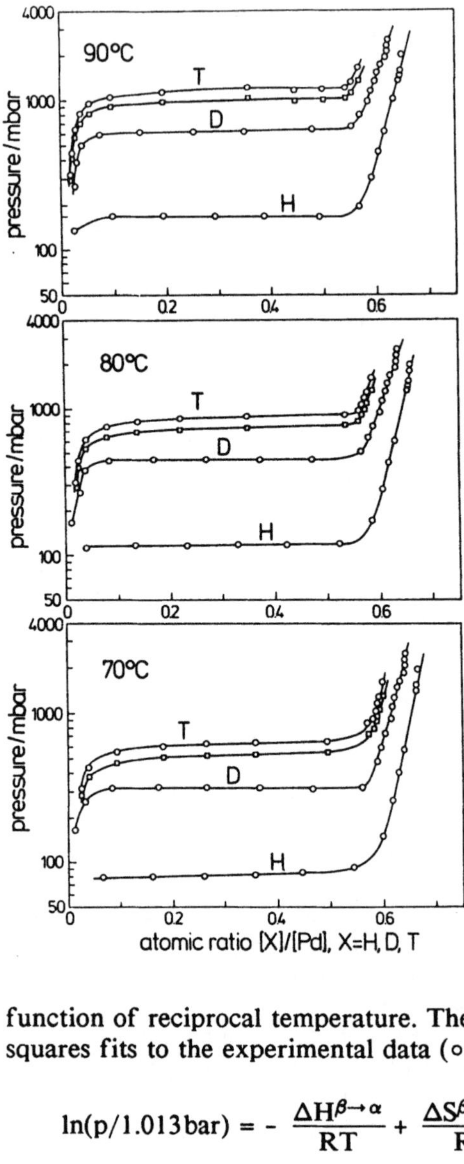

Fig.5.9. 70°, 80° and 90°C desorption isotherms [5.35] for pure hydrogen isotopes (o) protium (H), deuterium (D), tritium (T) and the isotope mixture (□) (95.2 at.% T, 1.2 at.% D, 3.7 at.% H) in Pd-X (X = H, D, T)

function of reciprocal temperature. The straight lines are the results of least squares fits to the experimental data (o) using the equation

$$\ln(p/1.013\,\mathrm{bar}) = -\frac{\Delta H^{\beta\to\alpha}}{RT} + \frac{\Delta S^{\beta\to\alpha}}{R} \tag{5.22}$$

where $\Delta H^{\beta\to\alpha}$ and $\Delta S^{\beta\to\alpha}$ represent the $\beta\to$a transition enthalpy and entropy of one mole X_2 gas, respectively. The $\Delta H^{\beta\to\alpha}$ and $\Delta S^{\beta\to\alpha}$ as well as the $\Delta H^{\alpha\to\beta}$ and $\Delta S^{\alpha\to\beta}$ values are listed in Table 5.3. In the case of T, the absorption isotherms were not determined because the uptake rate of tritium at higher concentrations was about 10 times lower than that of the stable hydrogen isotopes. This effect may probably be attributed to small amounts of ^{3}He in the tritium gas and/or other impurities due to the reactive nature of atomic tritium in the gas phase.

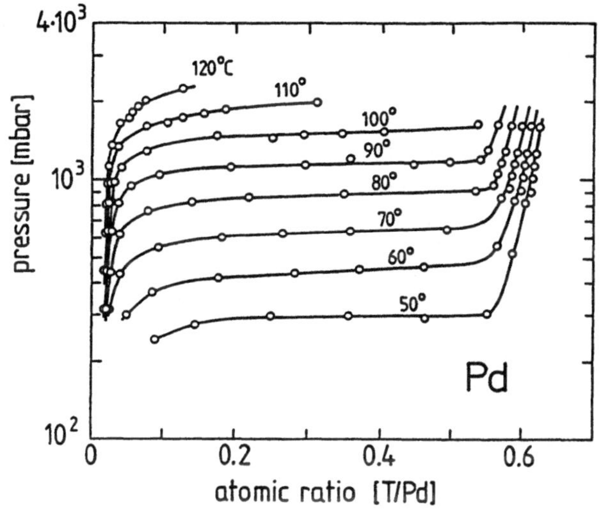

Fig.5.10. Desorption isotherms of the Pd-T system [5.39]

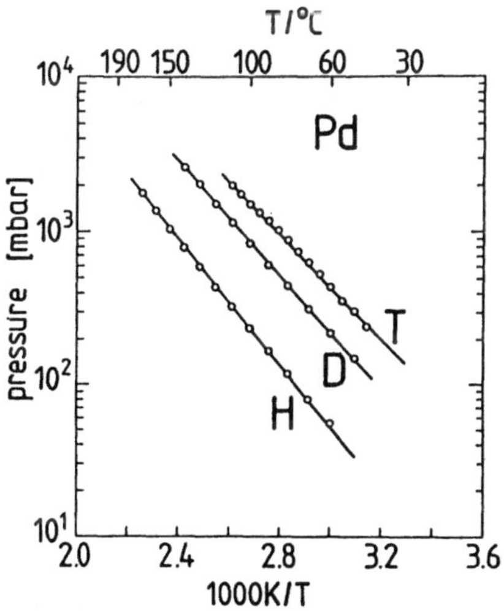

Fig.5.11. Desorption plateau pressures (o) of protium (H), deuterium (D) and tritium (T) in the miscibility gap of the Pd-X (X = H, D, T) systems as a function of reciprocal temperature [5.39]

Important publications [5.36, 37] conclusively discussing isotope effects have recently appeared. *Flanagan* et al. [5.36] derived equations which were compared with a limited number of experimental values. *Oates* et al. [5.37] calculated different physical properties of the Pd-X (X = H, D, T) system trying to get good agreement for many experimentally determined quantities. Their approach was to start from the known thermodynamic functions of the Pd-H system and to consider only the isotopic dependence of the optical modes. Since a satisfactory interpretation of the isotope effect at infinite dilution is not possible at present, they used the empirical

Table 5.3. $\beta \rightarrow \alpha$ ($\alpha \rightarrow \beta$) transition enthalpies $\Delta H_X^{\beta\rightarrow\alpha}$ ($\Delta H_X^{\alpha\rightarrow\beta}$) and entropies $\Delta S_X^{\beta\rightarrow\alpha}$ ($\Delta S_X^{\alpha\rightarrow\beta}$) of the Pd-X and $Pd_{0.90}Ag_{0.10}$-X (X = H, D, T) systems [5.22, 39]

System	$\Delta H_X^{\beta\rightarrow\alpha}$ [kJ/mol X_2]	$\Delta S_X^{\beta\rightarrow\alpha}$ [J/mol X_2 K]
Pd-H	39.0 ± 0.5	92.4 ± 1.3
Pd-D	35.4 ± 0.5	93.3 ± 1.3
Pd-T	32.3 ± 1.1	90.4 ± 3.2
$Pd_{0.90}Ag_{0.10}$-H	42.4 ± 1.0	94.7 ± 2.4
$Pd_{0.90}Ag_{0.10}$-D	38.9 ± 1.1	95.5 ± 2.9
$Pd_{0.90}Ag_{0.10}$-T	35.8 ± 0.9	92.0 ± 2.3
	$\Delta H_X^{\alpha\rightarrow\beta}$ [kJ/mol X_2]	$\Delta S_X^{\alpha\rightarrow\beta}$ [J/mol X_2K]
Pd-H	-37.4 ± 0.3	-92.5 ± 0.8
Pd-D	-33.6 ± 0.6	-93.3 ± 1.6

expression for the Sieverts' constant given by (5.16) and the parameter values of Table 5.2 for H, D and T in Pd. In addition, the excess chemical potential μ_H^E as determined by *Kuji* et al. [5.16] was employed. The main equation which allows one to determine any (p, r, T) triple for all three hydrogen isotopes in Pd is [5.37]:

$$RT \cdot \ln\sqrt{p_X} = RT \cdot \ln K_X^{\infty} + RT \cdot \ln\left(\frac{r}{1-r}\right) + \mu_X^E . \tag{5.23}$$

The excess chemical potential μ_X^E of the isotope X (X = D, T) can be calculated with the expressions [5.36, 37]

$$\mu_X^E = \mu_H^E - \mu_H^{E,V} + \mu_X^{E,V} , \tag{5.24}$$

$$\mu_X^{E,V} = \mu_X^V(r) - \mu_X^V(0) , \tag{5.25}$$

$$\mu_X^V(r) = \frac{3}{2}RC_X + 3RT \cdot \ln(1 - e^{-C_X/T}) + \frac{3}{2}Rr + \frac{re^{-C_X/T}}{1 - e^{-C_X/T}} , \tag{5.26}$$

$$C_H(r) = C_H^0(0) + Wr , \tag{5.27}$$

$$C_D = C_H/\sqrt{2} , \tag{5.28a}$$

$$C_T = C_H/\sqrt{3} . \tag{5.28b}$$

Table 5.4. Parameters a_0, a_1, b_0, b_1, c_0 and c_1 [5.37] used for calculation of the isotope independent part of the excess chemical potential: $\mu_H^E - \mu_H^{E,V} = (a_0+a_1T)r + (b_0+b_1T)r^2 + (c_0+c_1T)r^3$, see (5.24) and text].

a_0 [J/mol]	a_1 [J/mol K]	b_0 [J/mol]	b_1 [J/mol K]	c_0 [J/mol]	c_1 [J/mol K]
-67466.8	65.40	64634.49	-79.31	24355.39	-13.03

The isotope-independent contribution of the excess chemical potential (assuming only the optical modes to be isotope dependent) results from the difference of the first two terms on the right-hand side of (5.24). The addition of the vibrational contribution $\mu_X^{E,V}$ (X = D,T) to the isotope-independent excess chemical potential yields the total excess chemical potential μ_X^E (X = D,T). The vibrational excess chemical potentials for X = H,D,T are calculated with (5.25-28), where (5.26) describes the total vibrational contribution. Equation (5.27) represents the concentration dependence of the Einstein temperature C_X for X = H, D and T, where C_H^0 and W were determined from inelastic neutron scattering experiments [5.30,38]: C_H^0 = 800 K and W = -175 K. The parameters describing the isotope-independent part of the excess chemical potential are given in Table 5.4 [5.37].

The agreement of the results of (5.23) (plotted as solid lines) with the experimental data [5.22,23,39,40] is shown in Figs.5.12-14. The pressures are plotted as a function of concentration in the α and β phases and as a function of reciprocal temperature in the miscibility gap. In the α phase, the agreement is excellent. For the other phase areas, the data agree quite well considering the difficulties in the measurements of the concentrations due to the steep pressure rise in the β phase and the simple cubic representation of the isotope-independent contribution to μ_H^E (Table 5.4). In the two-phase areas, determination of the plateau pressures is complicated by hysteresis, but the difference between the log $p_{plateau}$ values are quite small.

In conclusion, the solubility data of Pd-D and Pd-T can be calculated accurately using an empirical expression for the Sieverts' constants, the excess chemical potential of H in Pd and the assumption that only the optical modes are isotope dependent [5.37].

5.2 Solubility of Protium, Deuterium and Tritium in Palladium Silver Alloys

Experiments on the solubility of tritium in $Pd_{1-z}Ag_z$ (z = 0.10, 0.20, 0.30) will be presented here together with the data for H and D obtained using

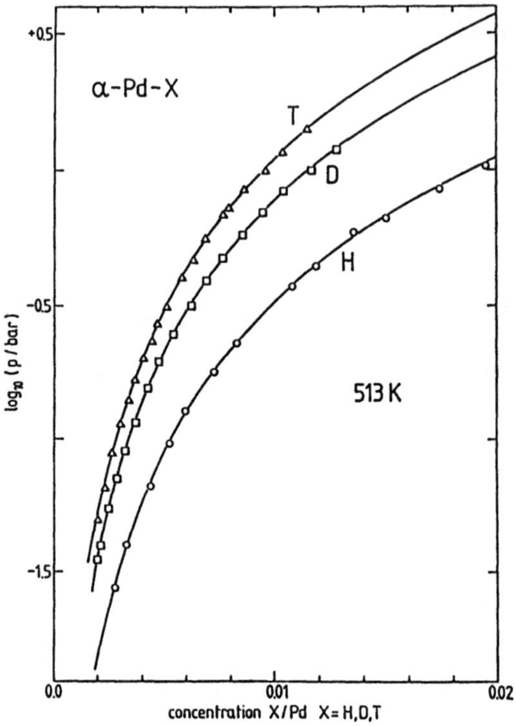

Fig.5.12. The 513 K isotherms (—) of protium (H), deuterium (D) and tritium (T) in the α phase of the Pd-X (X = H, D, T) systems calculated with (5.23) [5.37] in comparison with experimental values (o, □, Δ) from [5.23]

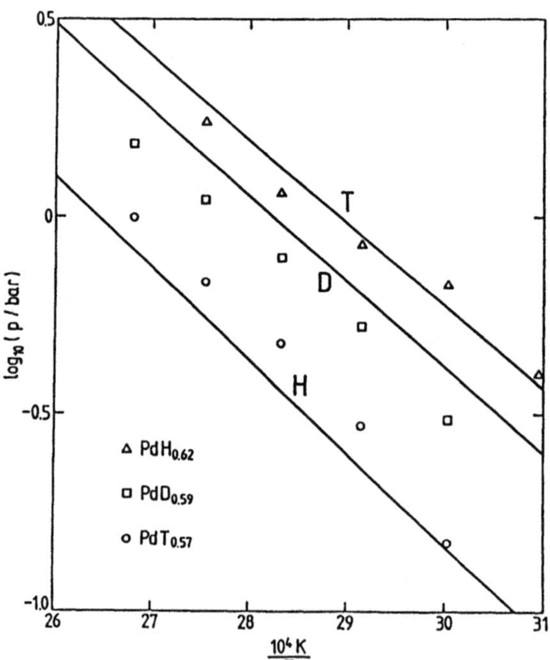

Fig.5.13. Equilibrium pressures (-) of protium (H), deuterium (D) and tritium (T) in the β phase of the Pd-X (X = H, D, T) systems calculated using (5.23) [5.37] in comparison with experimental values (o, □ [5.22], Δ: [5.39]

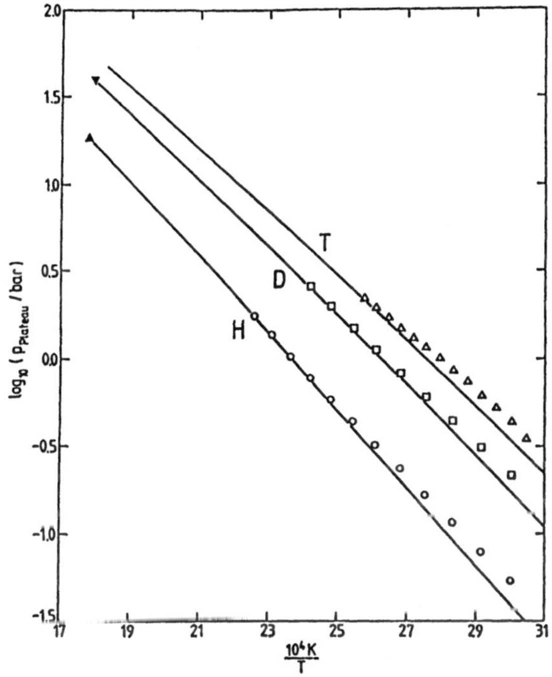

Fig.5.14. Plateau pressures (—) of protium (H), deuterium (D) and tritium (T) in the miscibility gap of the Pd-X (X = H,D,T) systems calculated using (5.23) [5.37] in comparison with experimental values (o,□) [5.22], (Δ) [5.39]; (▼▲) [5.40]

the same sample and the same experimental apparatus. A comparison with the data of other groups [5.41-51] will be made.

5.2.1 Solubility of Protium, Deuterium and Tritium in Palladium Silver Alloys for Hydrogen Concentrations $r \leq 0.02$

The 280°C desorption isotherms of $Pd_{0.90}Ag_{0.10}$ are plotted in Fig.5.15 together with the pure Pd data. In addition, the solubility data measured by *Boureau* et al. [5.46] and by *Blaurock* [5.50] are shown. Excellent agreement is noted considering the slightly higher temperatures used by *Boureau* et al. [5.46]. The solid lines in Fig.5.15 are the result of a least-squares fit through the experimental points using (5.5). These fits yield the Sieverts' constants which are presented for H, D and T in Figs.5.16-18. The data shown as squares and circles were determined independently by *Lässer* and *Powell*, respectively [5.52]. The dependence of the Sieverts' constants on the hydrogen isotope and on the Ag content can be described by the following inequalities:

$$K_H^\infty < K_D^\infty < K_T^\infty \tag{5.29}$$

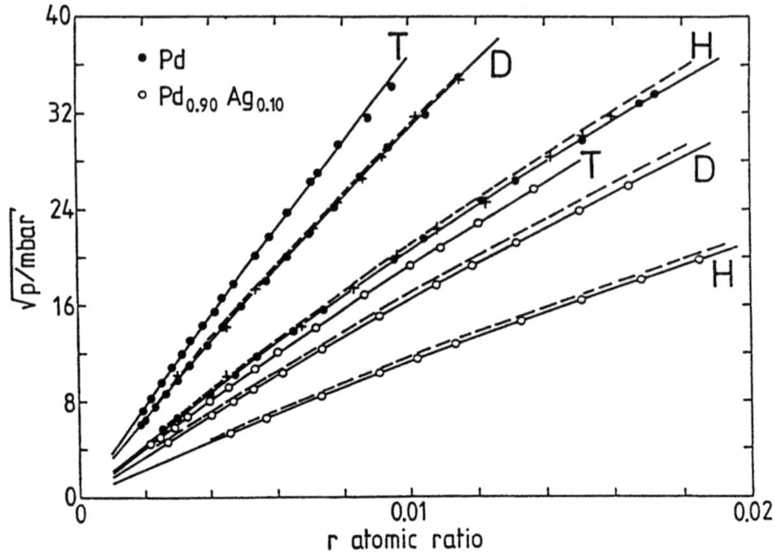

Fig.5.15. 280°C desorption isotherms of protium (H), deuterium (D) and tritium (T) in Pd (•) [5.23] and $Pd_{0.90}Ag_{0.10}$ (o) [5.39]. --: data measured at 282°C [5.46]; +: data measured at 280°C [5.50]

for the $Pd_{1-z}Ag_z$ alloys with z = 0.1, 0.2, 0.3 and

$$K_X^{\infty}(Pd) > K_X^{\infty}(Pd_{0.90}Ag_{0.10}) > K_X^{\infty}(Pd_{0.80}Ag_{0.20}) > K_X^{\infty}(Pd_{0.70}Ag_{0.30}) \tag{5.30}$$

for X = H,D,T at the same temperature.

In the case of alloys (e.g., $Pd_{1-z}Ag_z$) the physical description of the Sieverts' constants for H, D and T is even more difficult than for pure metals. The possible interstitial sites are no longer equivalent in the alloy. Given the fact that even in the case of Pd there is no quantitative description of the isotope effect at infinite dilution, the assumption was made that there is only one kind of hydrogen site in the $Pd_{1-z}Ag_z$ alloys and that the interaction between the alloy and hydrogen varies monotonically with the alloy composition. This again justified the use of (5.16). As in Sect.5.1 N = 1.0 and $C_I = 800/\sqrt{I}$ are used, where I = 1,2,3 for H,D,T. Assigning to A the value 2.0 and defining the parameters (E - Eo) and B as second-order polynomials in z, the following expressions [5.52] were determined:

$$E_0 - E = 28172.0 - \frac{1229.3}{\sqrt{I}} + \left(5796.4 - \frac{608.8}{\sqrt{I}}\right)z - \left(4387.8 + \frac{608.3}{\sqrt{I}}\right)z^2 \tag{5.31}$$

and

$$B = 238.6 + \frac{533.7}{\sqrt{I}} + \left(2493.9 + \frac{1920.4}{\sqrt{I}}\right)z + (24720.1 - 18362.0/\sqrt{I})z^2 \; . \tag{5.32}$$

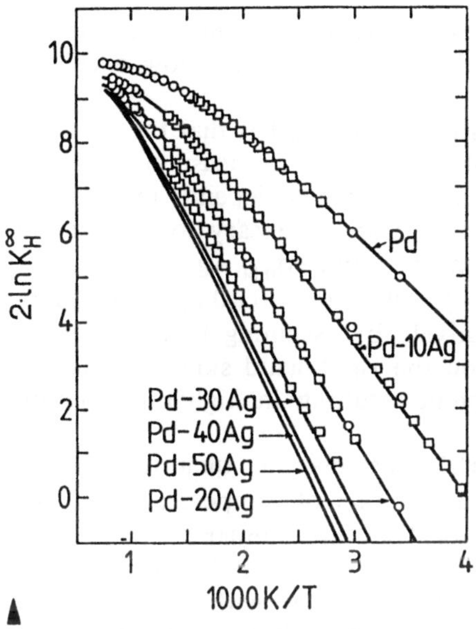

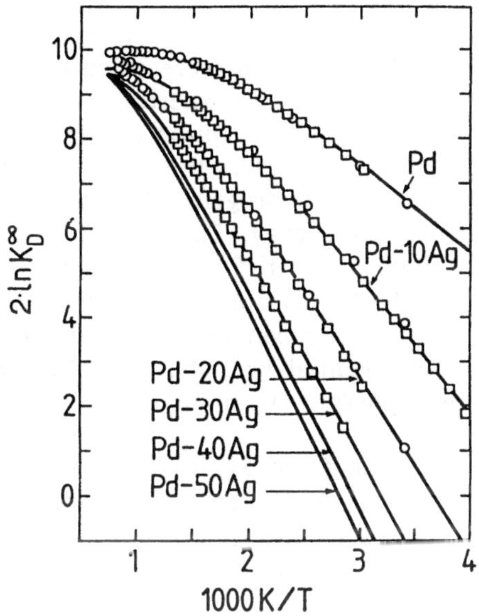

Fig.5.16. Sieverts' constant K_H^{∞} for protium in Pd and $Pd_{1-z}Ag_z$ (z = 0.10, 0.20, 0.30, 0.40, 0.50). □: data measured by *Lässer* [5.52], o: data measured by *Powell* [5.52]; solid lines: fit results obtained with (5.16,31,32)

Fig.5.17. Sieverts' constant K_D^{∞} for deuterium in Pd and $Pd_{1-z}Ag_z$ (z = 0.10, 0.20, 0.30, 0.40, 0.50). □: data measured by *Lässer* [5.52], o: data measured by *Powell* [5.52]; solid lines: fit results obtained with (5.16,31,32)

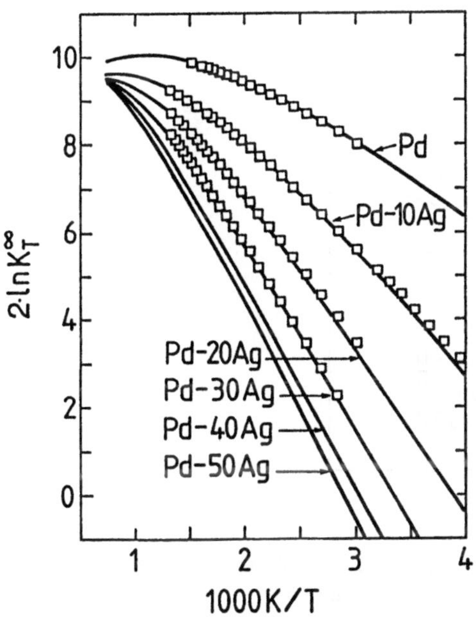

Fig.5.18. Sieverts' constant K_T^{∞} for tritium in Pd and $Pd_{1-z}Ag_z$ (z = 0.10, 0.20, 0.30, 0.40, 0.50). □: data measured by *Lässer* [5.52], o: data measured by *Powell* [5.52]; solid lines: fit results obtained with (5.16,31,32)

The usefulness of these empirical expressions is seen by the solid lines in Figs.5.16-18 which were calculated with (5.16,31,32) and show excellent agreement with the experimental data.

A simple description of the expression for $(E-E_0)$ is that the part independent of I gives the vibrational potential minimum measured with respect to the hydrogen atoms at rest. The part which depends on I is the zero point energy. The quadratic terms in z are necessary to predict the minimum of $\ln(K^{\infty})$ (solubility maximum) at about $Pd_{0.50}Ag_{0.50}$.

The higher solubility in the $Pd_{1-z}Ag_z$ alloys (z = 0.1,0.2,0.3) in comparison to Pd can be explained qualitatively by assuming that the vibrational potential well shown in Fig.5.6 on the right-hand side is shifted to lower energies with increasing Ag content resulting in higher dissolution enthalpies.

Using (5.16,31,32) for the Sieverts' constants, the quantities $\Delta\bar{H}_X^{\infty}$, $\Delta\bar{S}_X^{\infty}$, and $\Delta\bar{G}_X^{\infty}$ can be calculated as a function of temperature and Ag content. The results are not shown here. They are similar to the curves presented in Figs.5.4,5. The main difference is that for $z \leq 0.5$ the enthalpy curves are shifted to lower, more exothermic values with increasing Ag content. The $\Delta\bar{S}_X^{\infty}$ values do not show large changes because the dominant contribution stems from the entropy of the gaseous state.

The isotope effect in the solubility, $\log(K_X^{\infty}/K_H^{\infty})$ (X = D,T), is plotted against the reciprocal temperature in Fig.5.19. The ratios for X = D (which are more reliable than the data for X = T because for the latter

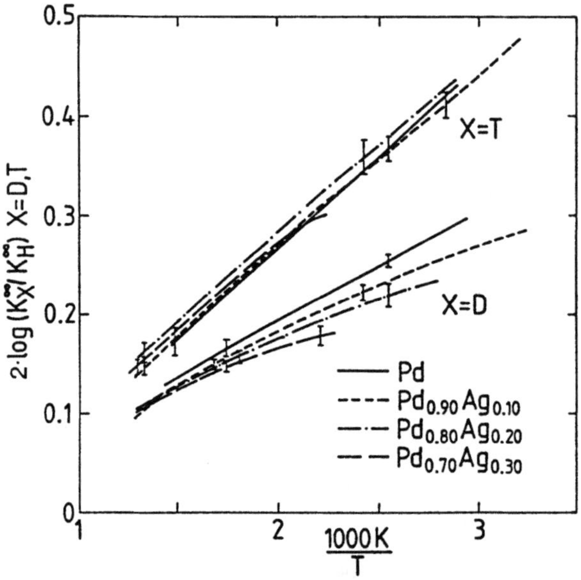

Fig.5.19. Ratio of the Sieverts' constants $(K_X^{\infty}/K_H^{\infty})$ (X = D,T) in $Pd_{1-z}Ag_z$ (z = 0.0, 0.10, 0.20, 0.30) as a function of reciprocal temperature [5.55]

extrapolation to 100% pure T_2 gas was necessary) depend weakly on the Ag content. In contrast, the absolute Sieverts' constants decrease strongly with increasing Ag concentration (Figs.5.16-18). The small decrease of the quantity log (K_X^∞/K_H^∞) may be explained by a small increase of the optical modes with increasing Ag content. This effect was also observed by *Chowdhury* and *Ross* [5.53] with inelastic neutron scattering in the β phases of $Pd_{1-z}Ag_z$-H (z = 0.10,0.20) alloys and by *Fratzl* et al. [5.54] for $Pd_{0.90}Ag_{0.10}$. As in the discussion in Sect.5.1, there exists no quantitative physical description of the isotope dependence of the solubility of $Pd_{1-z}Ag_z$-X (X = H, D, T) alloys at infinite dilution.

5.2.2 Solubility of Protium, Deuterium and Tritium in Palladium Silver Alloys for Hydrogen Concentrations $r \leq 0.50$

For the $Pd_{0.90}Ag_{0.10}$ alloy the 100° and 120°C desorption isotherms at higher concentrations are shown for H, D and T in Fig.5.20. In the miscibility gap below the critical temperature T_c^H = 173°C [5.42,45) the desorption plateau pressures are drawn in Fig.5.21 for the critical concentration r_c = 0.237 [5.45]. The values of the $\beta \rightarrow \alpha$ transition enthalpies (Table 5.3) show a strong isotope dependence similar to those for Pd. For all concentrations and all $Pd_{1-z}Ag_z$ alloys studied, the equilibrium pressures follow the inequality

$$p_{H_2} < p_{D_2} < p_{T_2} \tag{5.33}$$

at constant temperature and concentration.

A comparison of a part of H solubility values [5.55], presented as circles (o), squares (□) and triangles (Δ, ∇) for Pd, $Pd_{0.90}Ag_{0.10}$, $Pd_{0.80}Ag_{0.20}$ and $Pd_{0.70}Ag_{0.30}$, respectively, with the results of other groups is given in Fig.5.22 where the data of *Brodowsky* and *Poeschel* [5.41], *Gallagher* and *Oates* [5.44], and *Buck* [5.42] are shown using dotted, dot-dashed and dashed curves. With the exception of the data of [5.44] for $Pd_{0.90}Ag_{0.10}$, the agreement is very good considering that the Ag contents were not identical. As an example, for the tritium solubility in the $Pd_{1-z}Ag_z$ alloys studied, the desorption isotherms of T in $Pd_{0.80}Ag_{0.20}$ are shown in Fig.5.23. Furthermore, the 80° C desorption isotherms of T are drawn in Fig.5.24 for all $Pd_{1-z}Ag_z$ alloys studied. It is apparent that the behaviour previously observed for H [5.41,43,44] is also found in the case of T. Addition of Ag favours the absorption of T at lower equilibrium pressures and reduces the maximum T solubility at higher pressures as the Ag content increases. The physical reasons are mainly the following: the dilation causes an easier uptake of tritium atoms. The shrinkage and ultimately the disappearance of the plateau region is caused by the gradual filling of the Pd d-band by Ag electrons. *Oates* and *Ramanathan* [5.56] presented the first model discussing the role of configurational entropy due to different local environments in a statistically disordered alloy and of different site

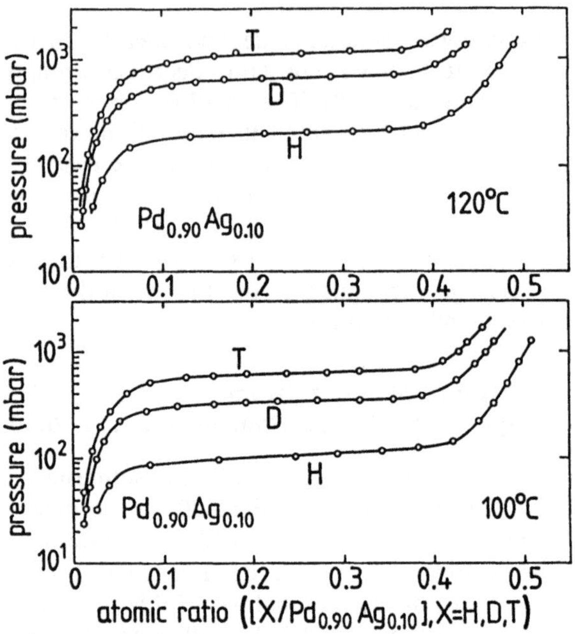

Fig.5.20. 100⁰ and 120⁰ C desorption isotherms for protium (H), deuterium (D) and tritium (T) in $Pd_{0.90}Ag_{0.10}$ [5.55]

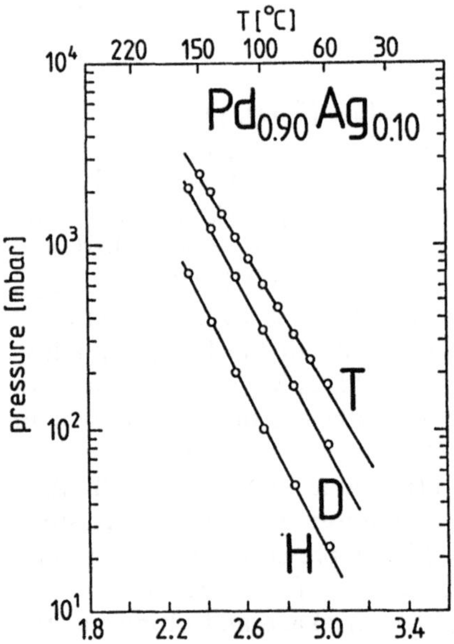

Fig.5.21. Desorption plateau pressures of protium (H), deuterium (D) and tritium (T) in the miscibility gap of the $Pd_{0.90}Ag_{0.10}$-X (X = H, D, T) systems as a function of reciprocal temperature [5.39]

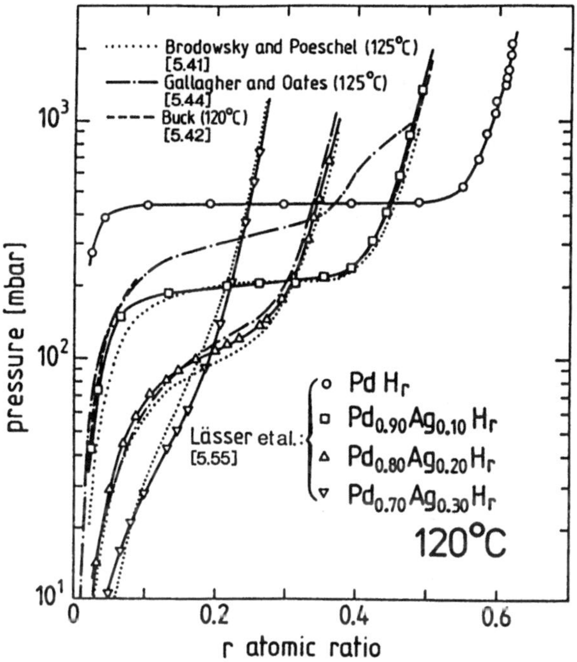

Fig.5.22. 120°C desorption isotherms of protium (H) in the $Pd_{1-z}Ag_z$ (z = 0.0, 0.10, 0.20, 0.30)-H systems [5.55] in comparison with the data of other groups (···: [5.41]; -·-: [5.44]; ---: [5.42])

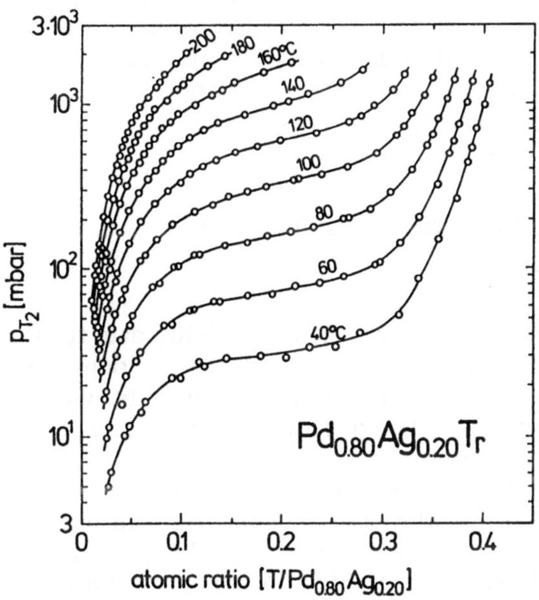

Fig.5.23. Desorption isotherms of the $Pd_{0.80}Ag_{0.20}$-T system [5.55]

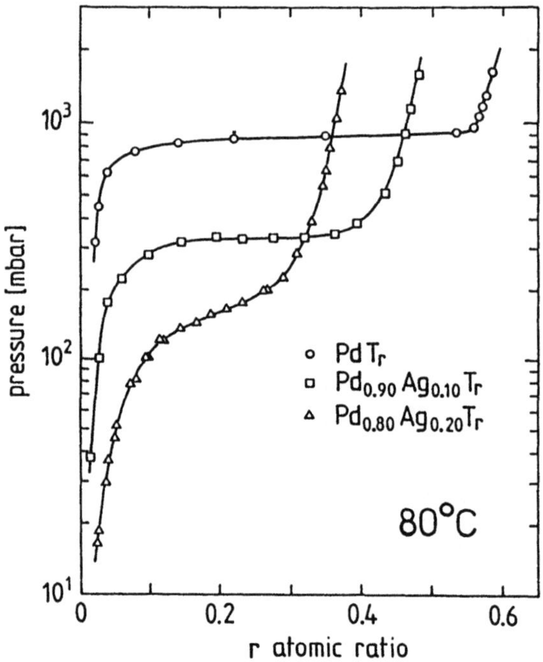

Fig.5.24. 80°C desorption isotherms of T in the $Pd_{1-z}Ag_z$ (z = 0.0, 0.10, 0.20)-T systems [5.55]

energies. Similar ideas were published by *Boureau* et al. [5.46]. Very recently, *Griessen* [5.57] gave a statistical description of hydrogen in a disordered material within the mean field approximation. This model reproduces the trends of the pressure-concentration-temperature relationship and of many Pd-based alloys such as $Pd_{1-z}Ag_z$ fairly well.

5.3 Solubility of Protium, Deuterium and Tritium in Lithium and Yttrium

Li and Y are discussed here because for these metals solubility data for the three hydrogen isotopes have recently become available. In addition, these two metals are of technological significance. Liquid Li may be used as breeding material and as a coolant in future fusion reactors, and Y has been proposed as a tritium storage material.

In the temperature range between 600° and 850°C and for pressures between 4 and 610 mbar, *Veleckis* [5.58] measured desorption plateau pressures of the $\alpha+\beta$ phase regions of the Li-X (X = H, D, T) systems. The desorption pressures versus 1/T are shown in Fig.5.25. The pressures to the left of the arrows belong to the $\alpha(1)+\beta(1)$ phases and those on the right side

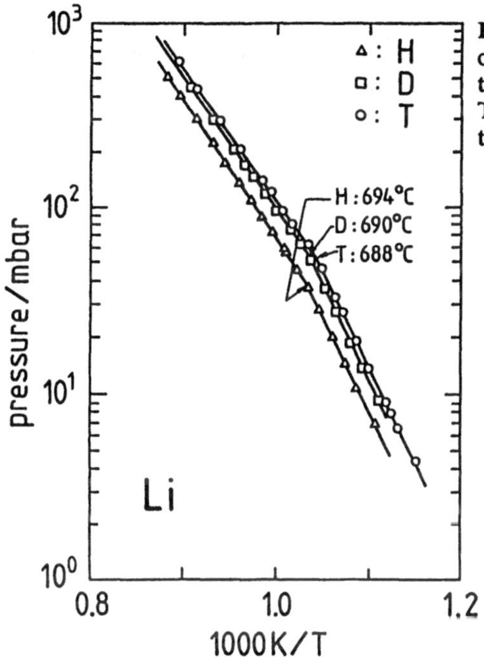

Fig.5.25. Desorption plateau pressures of protium (H), deuterium (D) and tritium (T) for the Li-(Li-X) (X = H, D, T) systems [5.58]. In addition, the monotectic temperatures [5.58] are indicated

to the $\alpha(\mathrm{l})+\beta(\mathrm{s})$ phases (l stands for liquid and s for solid). For each isotope the curves are approximated by two straight lines which intersect at the monotectic temperatures of the Li-(Li-X) (X = H, D, T) systems. The fitting parameters A and B of the straight lines

$$\ln(p[\mathrm{bar}]) = A - B/T \qquad (5.34)$$

and the monotectic temperatures [5.58] are listed in Table 5.5.

The plateau pressures and the monotectic temperatures both display a strong isotope effect: the equilibrium pressure increases at constant temperature in the sequence H, D, T, whereas the monotectic temperatures decrease.

Table 5.5. Monotectic temperature [5.58] and fit parameters A, B [5.58] obtained by a least squares procedure for the plateau pressures of the Li-LiX (X = H, D, T) systems.

System	Monotectic temperature [K]	ln(p[bar]) = A - B/T below monotectic A	below monotectic B [10^3K]	above monotectic A	above monotectic B [10^3K]
Li-LiH	967	20.88	23.38	14.72	17.42
Li-LiD	963	21.51	23.59	14.53	16.87
Li-LiT	961	21.25	23.21	14.30	16.53

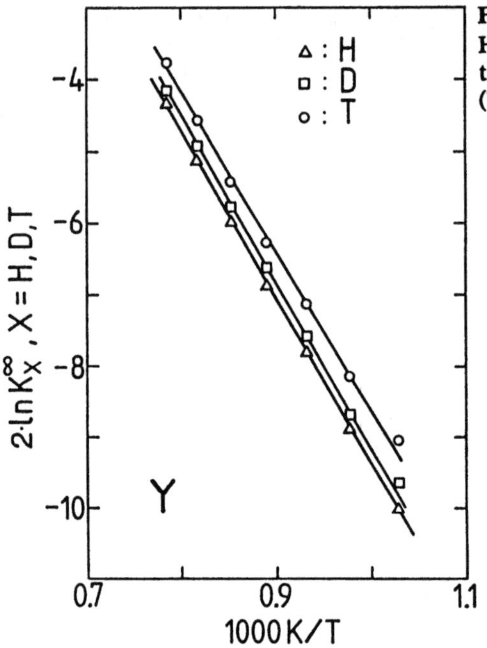

Fig.5.26. Sieverts' constant K_X^{∞}(X = H, D, T) for the hydrogen isotopes protium (H), deuterium (D) and tritium (T) in Y [5.59]

Veleckis [5.58] also calculated the standard free energies of formation yielding the following expressions:

LiH (s): $G_f/(\mathrm{kJ/mol}) = 82.73\ 10^{-3}\ T - 94.75\ ,$ (5.35a)
LiD (s): $G_f/(\mathrm{kJ/mol}) = 84.99\ 10^{-3}\ T - 95.17\ ,$ (5.35b)
LiT (s): $G_f/(\mathrm{kJ/mol}) = 86.42\ 10^{-3}\ T - 95.54\ .$ (5.35c)

These equations show that the formation of LiT is less exothermic than the formation of LiH and LiD.

In the range 700°-1000°C and for atomic ratios up to $r \simeq 2$, the equilibrium pressures of the Y-X (X = H, D, T) systems were measured by *Begun* et al. [5.59]. For concentrations $r \lesssim 0.3$ Sieverts' law is fulfilled. The Sieverts' constants are plotted in Fig.5.26 versus reciprocal temperatures. The Sieverts' constant for T is larger than those of D and H at constant temperatures. This means that the equilibrium pressure of T_2 is larger than that of D_2 and H_2 at equal concentration. *Begun* et al. claim that the isotope effect can be explained by considering the H atoms as three dimensional harmonic oscillators and using the vibrational energies determined by *Rush* et al. [5.60].

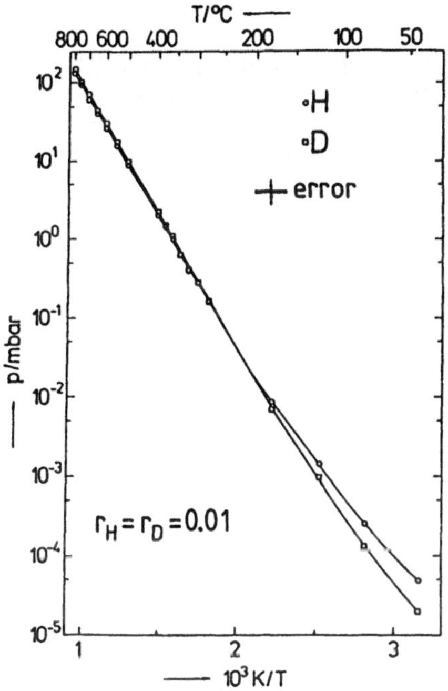

Fig.5.27. Equilibrium pressures for protium (H) and deuterium (D) in V for equal concentrations as a function of the reciprocal temperature [5.61]

5.4 Solubility of Protium and Deuterium in Vanadium

The observed isotope dependence of the solubility in the metals and alloys presented in the preceding sections may be described by (5.1,2). To avoid the impression that this behaviour may be found in every material reacting with hydrogen, the equilibrium pressures for the same H and D concentration ($r_D = r_H = 0.01$) in V are shown in Fig.5.27 versus reciprocal temperature. At temperatures below 220°C the equilibrium pressures [5.61] of deuterium are lower than those of protium. At temperatures above about 220°C the opposite effect is observed. The $\ln K_X^\infty$ (X = H, D) values [5.62] are depicted in Fig.5.28. They are in good agreement with the values determined by other groups [5.63,64]. Thus, the isotope effect in the very dilute α phase of V changes as a function of temperature yielding

$$K_H^\infty > K_D^\infty \quad T < 220°C \qquad \text{and} \tag{5.36a}$$

$$K_H^\infty < K_D^\infty \quad T > 220°C\,. \tag{5.36b}$$

The experimental values are fairly well described by a single straight line resulting in a first approximation in temperature-independent ΔH_X^∞ and $\Delta \bar{S}_X^\infty$ values (X = H, D).

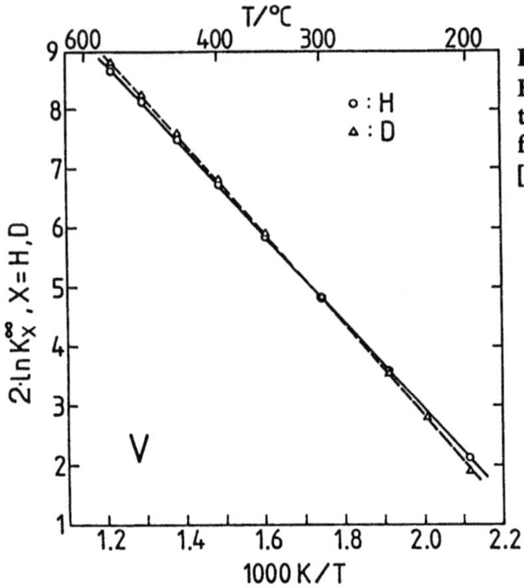

Fig.5.28. Sieverts' constant K_X^∞(X = H,D) for the hydrogen isotopes protium (H) and deuterium (D) in V as a function of the reciprocal temperature [5.62]

The ratio of the Sieverts' constants is plotted in Fig.5.29 [5.61] as a function of temperature. The solid line is obtained within the harmonic model. In V the tetrahedral sites are slightly distorted causing a splitting of the vibrational energies. Thus, instead of (5.14) the partition function ξ referenced to the gound state is given by

$$\xi = (1 - e^{-C_1/T})^{-1} (1 - e^{-C_2/T})^{-2} \tag{5.37}$$

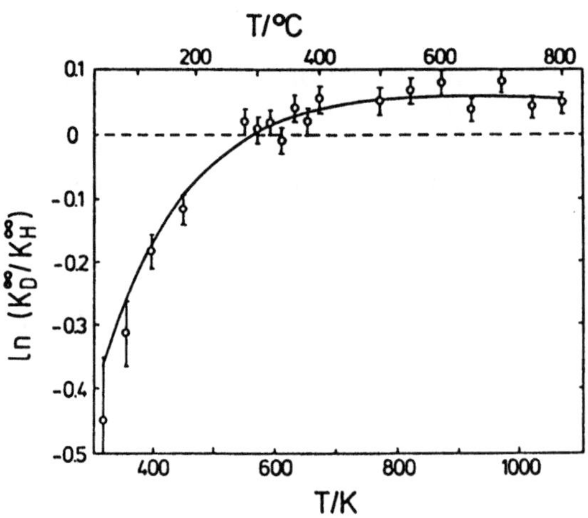

Fig.5.29. The ratio of the Sieverts' constant (K_D^∞/K_H^∞) in V as a function of temperature [5.61]

where C_2 is the "Einstein" temperature corresponding to the doubly degenerate vibrational ground state in V. The solid line in Fig.5.29 is calculated using (5.16) with the values $C_1{}^H = 1392.5$ K and $C_2{}^H = 2088.8$ K. These were determined by inelastic neutron scattering experiments [5.65]. In the case of V, the isotope effect in the ratio of the Sieverts' constants is well described within a simple harmonic model [5.61]. There are no reliable solubility data for T in V at low concentrations as yet. For higher concentrations, see Sect.3.6.4.

The isotope effect can be described qualitatively by referring to the illustrations in Fig.5.6. Given the fact that hydrogen atoms on tetrahedral sites in the bcc lattice of V have less space than on octahedral sites in the fcc lattice of Pd, the potential well of H atoms in V (left side) is steeper than the potential well in Pd. At low temperatures only the ground states in the metal are occupied. Here, more energy is gained by the absorption of D atoms from the gas phase than of H atoms. This is demonstrated qualitatively by the vertical lines. Therefore, a higher solubility is observed for D than for H. With increasing temperature a visual comparison is difficult since numerous states may be populated due to the degeneracy of the optical modes. For the sake of simplicity in Fig.5.6 the vibrational ground state levels are shown triply degenerate.

5.5 Isotope Dependence of the Phase Boundaries in the Pd-X, V-X and Nb-X (X = H, D, T) Systems

The absorption of hydrogen in metals can lead to the formation of various metal-hydrogen phases. These phases may differ in the arrangement of the dissolved hydrogen atoms. They may be distributed statistically or, alternatively, may be arranged on special sites in the host lattice. The various hydrides can be distinguished by the lattice parameter, the type of lattice structure and electronic properties. In the case of binary metal-systems, two different types of phase equilibria exist: phase separation and structural transformation. In the first case, two phases with identical structure are separated by a miscibility gap. The lattice parameter varies smoothly from one phase to the other as a function of hydrogen concentration. In the second case, the lattice structures of the phases are different.

The phase diagrams of many M-H systems are well known and have been published in books [5.66] and review articles [5.67,68]. The diagrams are given as a function of temperature and protium concentration. Widely differing experimental methods (lattice parameter studies, differential thermal analysis, resistivity, nuclear magnetic resonance, magnetic susceptibility, electron microscopy, etc.) have been employed to obtain reliable phase diagrams. A common feature of metal-hydrogen phase diagrams is that at high temperatures the hydrogen atoms are distributed statistically on the interstitial sites. At lower temperatures phase separation and order-disorder transitions may occur. An intriguing advantage of the study of metal-hydrogen phase diagrams when compared with metal-metal alloys is that, due to the high mobility of the hydrogen atoms, the phase transitions can be studied in thermodynamic equilibrium even at temperatures below 150 K. In reality, in this temperature range the metal-hydrogen systems are pseudo-binary alloys since only the hydrogen atoms are considered to be mobile in the host lattice (the concept of a lattice gas [5.69]). The metal atoms, however, are frozen in position.

In comparison to the many M-H phase diagrams presented to date, only a few studies on M-D phase diagrams have appeared. Very little is known about the phase boundaries of M-T systems.

In this section, the phase boundaries of T in Pd, V, and Nb will be compared with those for H and D. Tritium phase boundaries were published for Pd in [5.22,70,71], for V in [5.72,73,74,75] and for Nb in [5.76].

5.5.1 Phase Boundaries of the Pd-H, Pd-D and Pd-T Systems

Above 50 K the phase diagrams of the Pd-H and Pd-D systems contain an α phase with a low hydrogen concentration and a β phase with a high hydrogen concentration. These are separated by a miscibility gap. Below 50 K ordered metal-hydrogen phases have been found [5.77-79]. They will not be discussed here.

Due to the catalytic properties of the Pd surface for the dissociation and recombination of hydrogen molecules, handling of high concentration PdT_r samples outside the tritium loading apparatus is hazardous. Therefore, methods are used which allow the phase transitions to be determined from pressure (p), concentration (r) and temperature (T) in equilibrium measurements [5.70] performed with the tritium loading equipment [5.22,80]. A large Pd sample ($m_{Pd} \simeq 8.9$g) was charged such that the starting concentration of the PdX_r sample is just below the solvus in the two-phase region. The PdX_r (X = H,D,T) sample was heated in a small volume ($\simeq$71 cm^3) in temperature steps of about 3 Kelvin. During these quasi-isochoric desorption measurements, the concentration of the PdX_r sample had to cross the solvus between the α+β and α regions. Due to the small volume and the large Pd mass used, the concentration decrease upon heating is very small. Thus, absolute concentrations can be accurately determined.

Typical experimental desorption data for H, D, and T in Pd [5.70] are plotted in Fig.5.30. The arrows indicate the change of the slope in the curves drawn through the data and represent the intersection with the solvus line. Interpolation between the circles in the neighbourhood of the arrows gives the true temperature, concentration and pressure values of the phase boundary between the α+β and α phase as determined during desorption. The (T,r) data of the solvus line for H, D and T determined using other values of the starting concentration and the same experimental conditions are plotted as circles on the left-hand side of Fig.5.31. The cir-

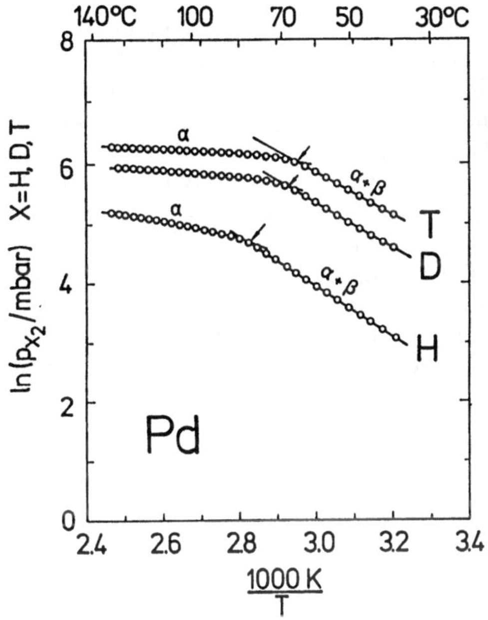

Fig.5.30. Desorption pressures for PdH_r, PdD_r and PdT_r samples as a function of reciprocal temperature [5.70]. The arrows indicate the change of slope due to crossing of the phase limit between the $\alpha+\beta$ and α phase areas

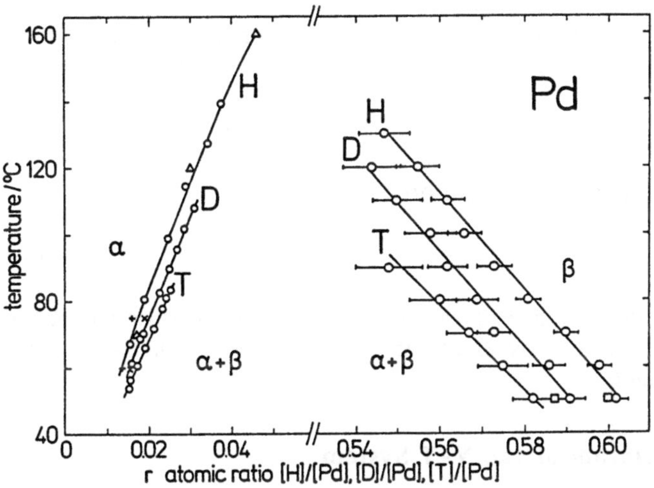

Fig.5.31. Part of the phase diagram of the Pd-H, Pd-D and Pd-T systems. o: [5.70]; Δ: PdH_r: [5.1]; +: PdH_r: [5.7]; x: PdD_r: [5.7]; □: PdH_r, PdD_r: [5.7]

cles on the right-hand side show the phase boundary values between the $\alpha+\beta$ and β regions. These data were also determined from desorption data (Sect.5.1 and [5.22]). The accuracy of these data is limited by the extrapolation procedure used to obtain the values from the isotherms [5.22]. For the circles on the left, the error bar is about equal to the circle diameter.

The curves through the circles in Fig.5.31 are plotted as a guide for the eye. In addition, a few data points determined by other groups [5.1,7] are also shown. The overall agreement is satisfactory. In the case of tritium, the phase boundaries could be determined only in a very limited range because the equilibrium pressures of H_2, D_2 and T_2 in contact with Pd obey the inequality (5.2) (Sect.5.1). Furthermore, only limited activities of below $3.7 \cdot 10^{13}$ Bq (1000Ci) were available. But the following conclusions can nonetheless be drawn: The solvus between the α phase and the miscibility gap is shifted to higher concentrations with increasing hydrogen isotope mass. The phase boundary between the $\alpha+\beta$ and β phase is shifted to lower concentrations in the sequence H, D, T. Thus the following inequalities describe qualitatively the phase boundaries r_a (r_b) between the α (β) phase and the miscibility gap

$$r_a^H < r_a^D < r_a^T \qquad \text{and} \tag{5.38a}$$

$$r_b^H > r_b^D > r_b^T \tag{5.38b}$$

for the same temperatures.

In other words, the miscibility gap shrinks and the single-phase regions α and β widen in the sequence H, D, T. As a consequence, the critical temperature $T_c{}^T$ of the Pd-T system is expected to be lower than those of the Pd-H and Pd-D systems, following the inequality

$$T_c^H > T_c^D > T_c^T \,. \tag{5.38c}$$

In analogy, one expects for the critical pressures

$$p_c^H < p_c^D < p_c^T \,. \tag{5.38d}$$

Recently the isotope dependence of the phase boundaries mentioned above was calculated quantitatively by *Oates* et al. [5.37] (Sect.5.1). Satisfactory agreement was obtained.

5.5.2 The Phase Diagram of the V-T System

The systems V-H and V-D show the largest isotope dependence of metal hydride phase diagrams. The differences in the lattice structure, the phase boundaries and the statistical and ordered occupation of tetrahedral and/or octahedral sites by H and D were discussed in recent review articles [5.68,81,82]. The following methods were employed for the determination of the V-T diagram: 1) NMR technique [5.73], 2) Gorsky effect [5.74], 3) lattice parameter measurements [5.72], 4) differential thermoanalysis [5.75] and 5) autoradiography [5.75].

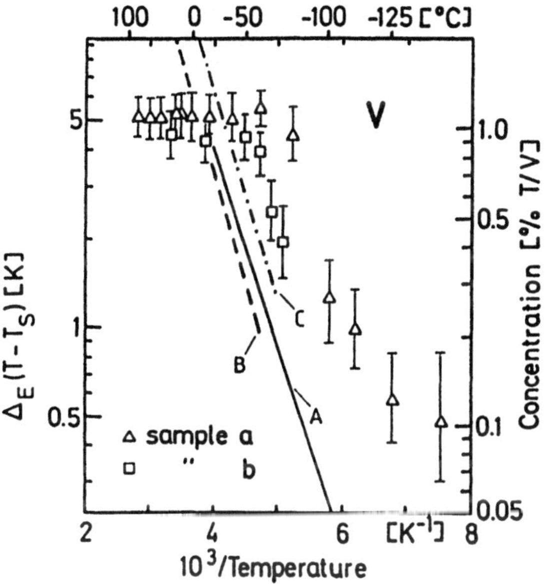

Fig.5.32. Product of the relaxation strength Δ_E and $(T-T_s)$ (left-hand side) and concentration (right-hand side) versus reciprocal temperature for two different samples a and b. [5.74]. Solubility limit for H in V: line *A* [5.85], line *B* [5.86]; solubility limit for D in V: line *C* [5.87]

The change of the relaxation strength Δ_E obtained by Gorsky effect measurements in the presence of precipitation allows a rough determination of the (T, r) values of the solvus line by means of the equation [5.83, 84]:

$$\Delta_E = A \frac{rP^2}{T - T_S}, \tag{5.39}$$

where A is a constant depending on the elastic coefficients [5.83], P is the trace of the dipole moment tensor and T_s the spinodal temperature.

The product of the relaxation strength Δ_E and $(T-T_s)$ is shown as a function of reciprocal temperature in Fig.5.32. On the right, the concentration r of the solvus line obtained with (5.39) is shown. In addition, the phase boundaries of H and D are indicated by straight lines measured by other groups [5.85-87]. In the temperature range above -50°C the product $\Delta_E(T-T_s)$ is constant because the concentration r of the T atoms in the α phase of V does not change. Below -50°C the product decreases due to precipitation of T atoms into the β phase, thus decreasing the maximum tritium concentration in the α phase. A comparison of the H, D and T solvus data shows that at constant temperature the concentration of the solvus line increases with increasing hydrogen isotope mass.

Above room temperature the solvus concentration of T in V was studied by lattice parameter determinations [5.72]. The results obtained upon

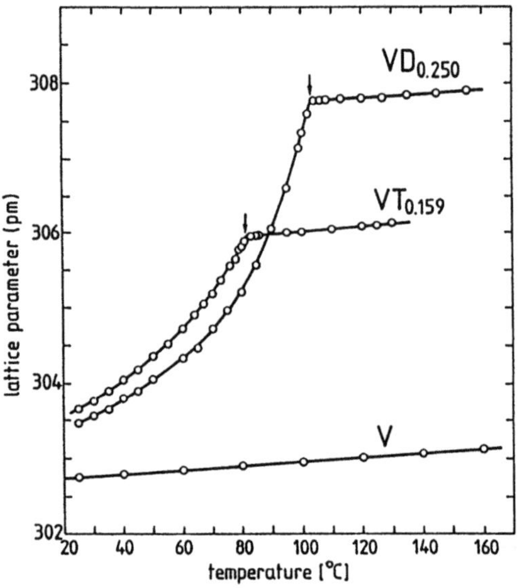

Fig.5.33. Lattice parameter of $VT_{0.159}$, $VD_{0.250}$ and pure V samples as a function of temperature [5.72]

heating are shown in Fig.5.33 as a function of temperature. The concentrations of the hydrogen-charged V samples were in the two-phase area at room temperature. In both samples, the lattice parameters corresponding to the phase limits between the α, $\alpha+\beta$ and $\alpha+\beta$, β phases may be measured. The lattice parameter on the left of the arrows in Fig.5.33 corresponds to the solvus line. The arrows indicate the temperatures at which the samples leave the two-phase region. The straight lines on the right show the temperature dependence of the lattice parameter of the sample $VD_{0.250}$ and $VT_{0.159}$ in the pure α phase. The lattice parameter of a pure V single crystal is also presented. The differences between the lattice parameters of $VT_{0.159}$ or $VD_{0.250}$ and the pure V (Fig.5.33), divided by the relative lattice-parameter change per atomic hydrogen concentration, yield the hydrogen-to-metal ratios plotted in Fig.5.34 as circles and squares, respectively. For the analysis of the $VD_{0.250}$ and $VT_{0.159}$ samples the same relative lattice-parameter change per atomic hydrogen concentration $\Delta a/a\Delta x = (0.0633\pm0.003)$ as given by *Peisl* [5.88] for VH_x was used, since no isotope dependence of the lattice expansion was noted for H and D in V [5.88]. In addition, the solvus line values listed in [5.89] are also plotted in Fig.5.34. These data were determined by means of differential thermal analysis (DTA). The agreement with the values measured with the Bond method [5.72] is not satisfactory. The reasons may be twofold: 1) in DTA experiments it is difficult to accurately determine the phase transition temperature from the shape of the DTA signal; 2) using X rays for measurements of the lattice parameter, only a few microns at the sample surface are analysed. The solvus concentration values obtained may be influenced by the oxide layer at the surface and/or oxygen atoms dissolved in the metal near the surface. The main conclusion to be drawn from Fig.5.34 is in agree-

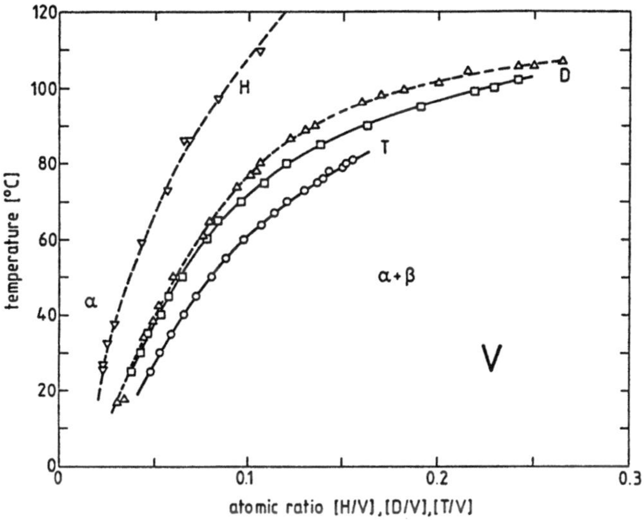

Fig.5.34. Solvus lines of the V-H, V-D and V-T systems. (o: VT_r [5.72]; □: VD_r [5.72], Δ: VD_r [5.89], ∇: VH_r [5.89])

ment with the observations in Fig.5.32: the solvus line is shifted to higher concentrations in the sequence H, D, T. Because only samples with tritium activities below $2.775 \cdot 10^{10}$ Bq (0.75Ci) could be studied on the X-ray diffractometer, single crystals with higher concentrations were not investigated with the Bond method.

For higher concentrations, the (T, r) values of the V-T phase diagram were determined with DTA because a commercial DTA 2000 equipment (Mettler) was permanently located in the control areas. To prevent tritium leakage from the Al crucibles due to internal pressure increases during heating and subsequent contamination of the DTA apparatus, the VT_r samples were cold-welded into the usual DTA-Al crucibles at an air pressure of less than half a bar. Examples of the DTA results are shown in Fig.5.35. The different phases and heating rates are indicated. The transition temperatures between different phase regions were determined from the intersections of the dashed lines.

Using all the information discussed above, the phase diagram of the V-T system was constructed as shown in Fig.5.36. The triangles Δ, ∇,◁, circles o and squares □ were determined by means of the Gorsky effect [5.74], the Bond method [5.72], autoradiography [5.75], DTA [5.75] and NMR [5.73] techniques, respectively. The solid lines describe well-characterized phase limits. The dashed lines are phase boundaries which were not directly determined using one of the above methods, but were extrapolated on the basis of the known behaviour of the V-D system. The phase limit value shown as ◁ represents a new application of tritium autoradiography for a direct observation of phase transitions in metal tritide systems. The autoradiographic picture of a VT_r sample with the concentration r of the

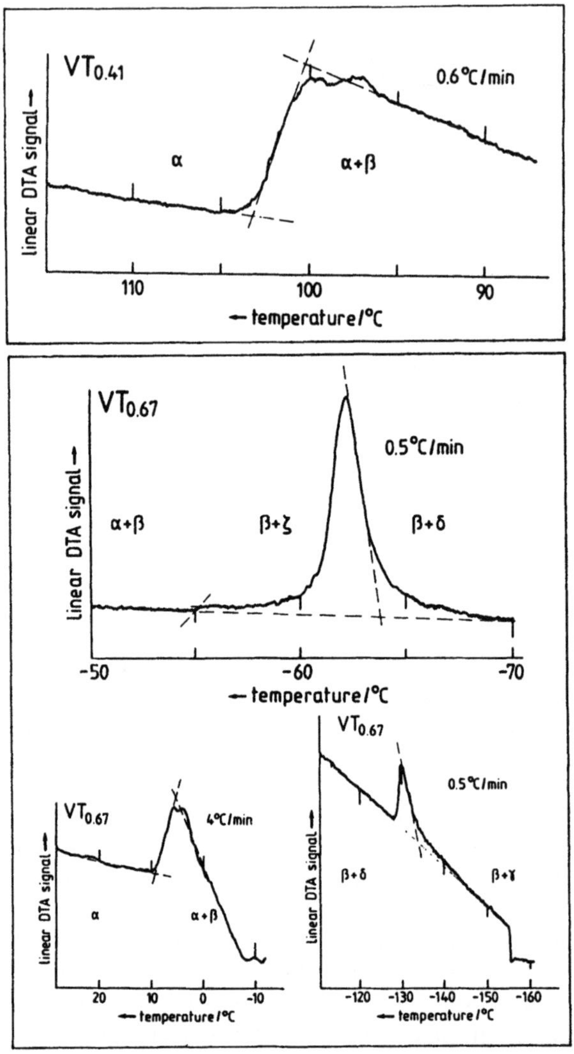

Fig.5.35. DTA signals of VT_r samples as a function of temperature [5.75]

two-phase region $\alpha+\beta$ shows the typical features presented in Fig.4.4. Upon heating, the number of needles which represent the β phase domains decreases. When the transition temperature to a single phase is reached, the contrast in the exposed X-ray sensitive film between phases, with different tritium concentration, disappears. The transition temperature of a $VT_{0.103}$ sample was found to be between 65° and 68°C. This was in good agreement with the lattice parameter data [5.72].

A comparison of the V-T system plotted in Fig.5.36 with the V-H and V-D systems drawn as dot-dashed and dotted lines is shown in

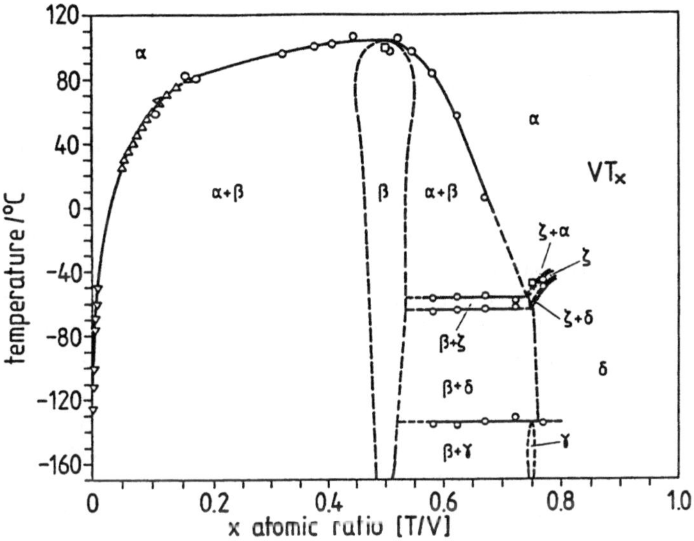

Fig.5.36. Phase diagram of the V-T system. (o: DTA results [5.75], ◁ : autoradiographic result [5.75], □: NMR results [5.73]; Δ: Gorsky effect [5.74] and ∇: lattice parameter measurements [5.72])

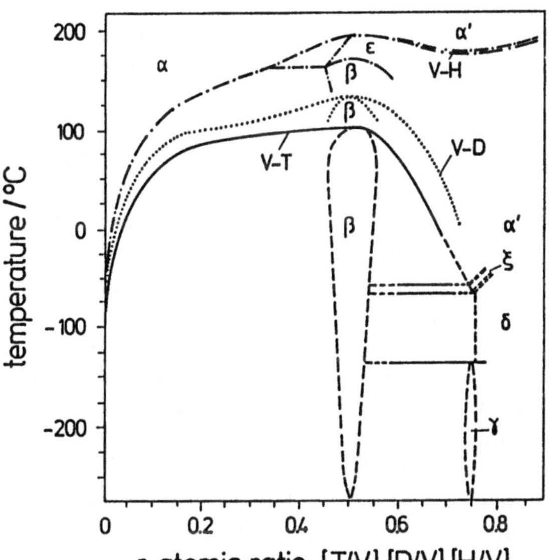

Fig.5.37. Phase diagram of the V-T system (solid lines) in comparison with those of the V-D (dots) and V-H (dot-dashed lines) systems [5.75]

Fig.5.37. The phase diagrams of the V-T and V-D systems agree topologically. The V-T and V-H on the one hand and the V-D and V-H phase diagrams on the other, are very different in the concentration range above r=0.33 as evidenced by the disappearance of the $\epsilon(\beta')$ phase in the case of the V-D and V-T systems. Most analogous phase boundaries occur at lower temperatures in the sequence H, D, T. A comparison of the V-D and V-T phase diagrams shows that for T the $\alpha+\beta \rightarrow \alpha$, $\beta \rightarrow \alpha$ and $\beta+\gamma \rightarrow \beta+\delta$ phase transitions are shifted to lower temperatures. The $\beta+\zeta$ phase limits are more or less the same for T and D in V.

To date, no theoretical description of the V-H, V-D and V-T phase diagrams or of the isotope dependences has been given. However, attempts [5.90] were made to discuss parts of the phase diagrams; for example, the solvus behaviour [5.89,91,92] has been discussed. The main origin of the isotope differences in the V-T and V-D phase diagrams is probably different vibrational energies of the T and D atoms on the tetrahedral sites in the α phase and on the octahedral sites in the β phase. Accurate inelastic neutron scattering experiments for the determination of the vibrational energies in the ground state and in excited states and of the vibrational potential for the different phases would be highly desirable [5.34].

It is hoped that the present V-H, V-D and V-T phase diagrams will stimulate further theoretical work on the isotope effect and on the phase limits.

5.5.3 The Phase Diagram of the Nb-T System

The phase diagram of the Nb-H system has been frequently reviewed [5.68,93,94] in recent years and is still open to question [5.82]. It can be divided into two parts - one above room temperature [5.95-97], which is well established in the concentration range below H/Nb=1 with the exception of the isotopic dependence of the critical temperature [5.98,99], and one below room temperature [5.100], which is complicated due to the presence of several different phases, and also somewhat controversial [5.82].

The phase diagram of the Nb-T system above room temperature will be compared in this section with the Nb-D and Nb-H diagrams. Below r=1, only three phases are known in this temperature range, the α and α' phases (where the hydrogen atoms are statistically distributed on tetrahedral sites in the bcc structure of Nb with different lattice parameters in the α and α' phases), and the β phase where a slight orthorhombic distortion of the original cubic Nb lattice occurs due to the ordering of hydrogen atoms.

The experimental methods used to determine the phase limits of the Nb-T system were X-ray diffraction and DTA. The resulting phase diagram is plotted in Fig.5.38 by solid lines. The Nb-H phase diagram is represented by dashed curves. From the known isotopic dependence of the Nb-H and Nb-D phase diagrams, parts of the Nb-T phase boundaries are extrapolated using a few experimentally determined phase boundary values of the Nb-T system as fixed points and assuming that the Nb-H, Nb-D and Nb-T phase diagrams are topologically equivalent. The extrapolated isotope dependence of the critical temperature for T is indicated in Fig.5.38. The dependence for H and D was determined by recent solubility

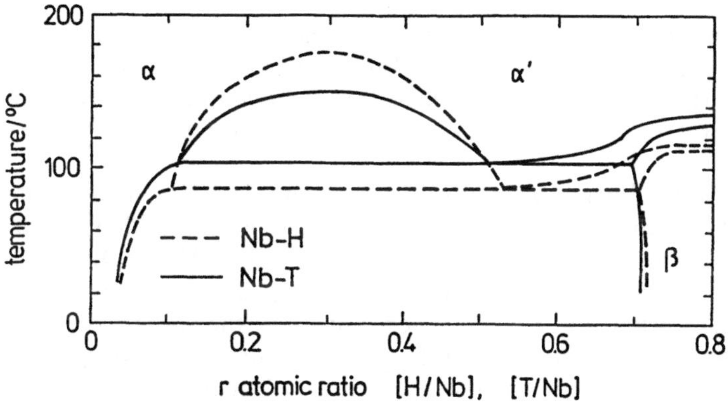

Fig.5.38. Phase diagram of the Nb-H (--) system [5.95,97,99] and Nb-T (—) system [5.76]

measurements [5.99]. These results are at variance with earlier work [5.98]. For clarity, the Nb-D phase diagram is omitted in Fig.5.38. The main isotope effects in the Nb-H, Nb-D and Nb-T phase diagrams are: 1) The triple line is shifted to higher temperatures with increasing hydrogen isotope mass. The monotectoid temperatures for H, D and T in Nb are 87.5°±1.0°C [5.76], 99°±1°C [5.96] and 103.9±1.0°C [5.76], respectively. 2) The limits between the α' and $\alpha'+\beta$ phases and 3) the boundaries between the β and $\alpha'+\beta$ phases, are shifted to higher temperatures in the sequence H, D, T. 4) The critical temperatures decrease: T_c^H=177°C, T_c^D=157°C [5.99]. 5) The maximum solubility in the α phase decreases with increasing isotope mass. The effect is strongest near the triple temperature, but also observable at low temperatures, as mentioned by *Zag* [5.101]. This isotope dependence of the solvus line in Nb is opposite to the behaviour discussed for Pd and V in the preceding sections. Very recently, isotope effects and hysteresis for the triple point transition in the Nb-X (X=H,D,T) systems were discussed by *Craft* et al. [5.102].

5.6 Optical Vibrations of Protium, Deuterium and Tritium in Niobium

Neutron spectroscopy has become a very powerful tool for studying metal-hydrogen systems. It offers the possibility of obtaining important information on the structure, occupied sites, diffusion coefficients, vibration energies, elastic constants, etc. The techniques have now been refined to such an extent that measurements of higher harmonics of localized vibrations of H and D atoms in metals are feasible [5.30,34,103-105]. Using neutron spallation sources, vibrational energies with quantum numbers as high as 14 have been observed [5.34]. The energies of the optical modes for H or D have already been determined in many metals. Due to the high incoherent cross section of H in comparison with D (Table 1.1) most inelastic neutron scattering experiments have been performed for H. In contrast, information about structures was in most cases obtained using metal deuteride samples (Table 1.1).

The first measurement of tritium vibrations [5.106] in a metal appeared in 1981, much later than for metal protide and deuteride systems. The metal studied was niobium. The vibration spectra of H and D in Nb are well known [5.103, 105]. The Nb sample was charged at 300°C with tritium to a concentration of $NbT_{0.2}$. According to the phase diagram (Sect.5.5.3), at room temperature the $NbT_{0.2}$ sample consists of two phases, the α and β phases, with tritium concentrations of about 3% and 71%, respectively. About 25% of the sample is in the β phase. The oxide layers on the surface of the $NbT_{0.2}$ sample can be considered as a first permeation barrier (Sect.4.2 and [5.76, 107]), and in addition, a molybdenum cylinder was used as a second containment for the $NbT_{0.2}$ sample. The activity of the $NbT_{0.2}$ sample was about 25 TBq ($\simeq$ 676Ci). Measurements of the optical modes of tritium are much more difficult, since the scattering cross section of T is much lower than that of H (Table 1.1), and a higher background is inevitable due to the secondary containment. This requires that the protium content of the T_2 gas employed is very low in order to prevent H peaks from dominating the tritium spectrum. Furthermore, ^3He is generated by tritium decay in the samples, therefore the neutron intensity will decrease with time due to the very large absorption cross section σ_a of ^3He ($\sigma_a \simeq 5327b$ for ^3He). The experiments should thus be performed as soon as possible after T charging.

Figure 5.39 shows the neutron spectra for $NbT_{0.2}$, $NbD_{0.725}$ and $NbH_{0.32}$ at 295 K [5.106]. The $NbD_{0.725}$ and $NbH_{0.32}$ spectra agree with published data [5.103, 105] demonstrating that as much as 67% α phase does not significantly affect the spectra attributed to the β phase. The two peaks of the $NbD_{0.725}$ and $NbH_{0.32}$ spectra can be assigned to the optical modes of the D and H atoms in the slightly distorted tetrahedral interstices of the orthohombic phase. The intensity of the peaks to the right is about twice that at lower energy because, due to the distortion, the higher vibrational modes are doubly degenerate. In the case of the $NbT_{0.2}$ spectrum, four well-resolved peaks are seen. The two dominant peaks in the centre are attributed to the optical vibrations of tritium atoms in the β phase of the Nb-T system. Again the energies of the T optical modes are shifted to lower values due to the larger mass of the T atoms. These shifts are expected from the well-known behaviour of H and D. They can be clearly seen in the spectra of $NbH_{0.32}$ and $NbD_{0.725}$. The small peak to the left of the $NbT_{0.2}$ spectrum is assigned to vibrations of oxygen atoms which diffused into the sample during the charging procedure. The amount of oxygen impurities is estimated to be about 1 at.%. The small peak to the right of the $NbT_{0.2}$ spectrum is attributed to H impurities absorbed from the slightly impure T_2 gas used for charging. Due to increasing hydrogen iso-

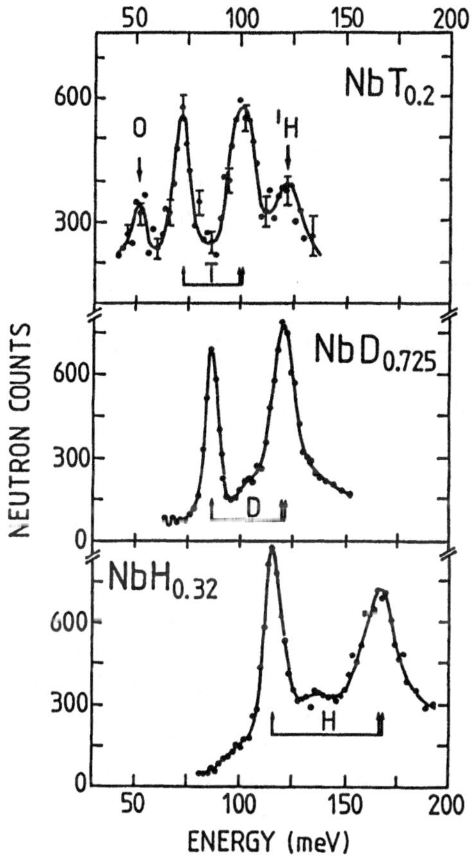

Fig.5.39. Inelastic neutron spectra of $NbT_{0.2}$, $NbD_{0.725}$ and $NbH_{0.32}$ at 295 K [5.106]

tope mass the energies of the H, D and T peaks are shifted to progressively lower values. Numerical values are given in Table 5.6. The vibrational energies of the different isotopes cannot be described by a simple harmonic model, since the ratios of the vibrational energies do not follow the inverse square root of the masses. Instead, anharmonic contributions to a parabolic potential have to be considered.

Table 5.6. Vibration energies ($\hbar\omega_1, \hbar\omega_2$) of protium (H), deuterium (D) and tritium (T) in β-phase Nb-X (X = H,D,T) [5.106]

	$\hbar\omega_1$ [meV]	$\hbar\omega_2$ [meV]
H	116 ± 0.7	167 ± 1.5
D	86 ± 1	120 ± 1.5
T	72 ± 1	101 ± 1

5.7 Superconductivity of the Pd-H, Pd-D and Pd-T Systems

The discovery of relatively high superconducting transition temperatures T_c^H in Th_4H_{15} by *Satterthwaite* and *Toepke* [5.108], in PdH_r with $r \geq 0.8$ by *Skoskiewicz* [5.109] and of the remarkable inverse isotope effect by *Stritzker* and *Buckel* [5.110] (i.e., a higher $T_c^D \simeq 11K$ in PdD in comparison to $T_c^H \simeq 9K$ in PdH) stimulated the interest in metal protides and deuterides. Although the first superconductivity studies of the Pd-H and Pd-D systems were published already 17 years ago, preliminary results for PdT_r were presented only recently.

Schirber et al. [5.111] measured the superconducting transition temperature of PdT_r samples inductively by means of a bridge technique. They mixed one-third of the Pd powder sample used with Sn, which worked as an internal thermometer. The Pd powder was charged at room temperature with tritium gas at pressures up to 700 bar and cooled to 77 K. After removal of the T_2 gas, the pressure vessel was filled with ^4He gas and immersed in a liquid ^4He bath, so that the ^4He gas inside the chamber improved the thermal contact between the sample and the vessel walls. A temperature increase inside the chamber is expected due to the heat of 0.324 W per g tritium caused by the decay of tritium. The transition temperatures were measured outside the pressure chamber. The observed superconducting transition temperatures of Sn in contact with PdT_r were shifted by 0.2 K to lower temperatures in comparison to the Sn transition at 3.72 K. This means that the PdT_r powder in contact with Sn is probably 0.2 K warmer than the outside wall of the pressure vessel. The PdT_r sample which was not mixed with Sn should be even warmer. Thus, the transition temperatures for PdT_r shown in Fig.5.40 were obtained by adding the observed 0.2 K shift. In view of these problems, the plotted

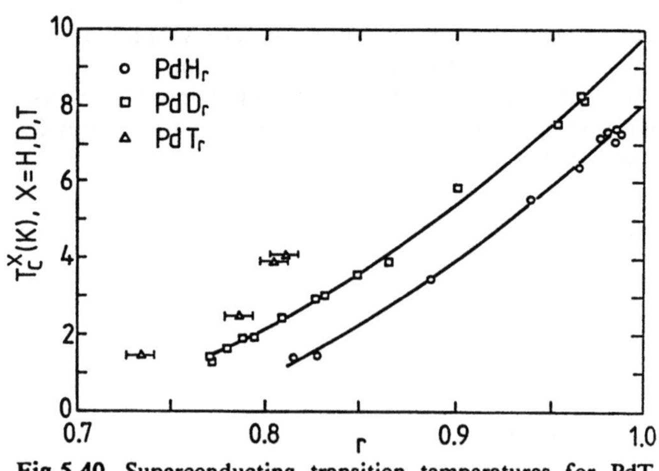

Fig.5.40. Superconducting transition temperatures for PdT_r [5.111] in comparison with the results for PdH_r and PdD_r [5.112] as a function of hydrogen concentration r

data represent only a lower bound to the transition temperature T_c^T in the Pd-T system. Figure 5.40 also shows the transition temperatures of the Pd-H and Pd-D systems measured by *Schirber* and *Northrup* [5.112]. In spite of the difficulties mentioned above, a further large increase of the transition temperature is observed, as expected from the known behaviour of H and D in Pd. The large isotopic effect is described by the inequality

$$T_c^H(r) < T_c^D(r) < T_c^T(r) , \quad r > 0.7 , \tag{5.40}$$

for equal concentrations r. This is called an inverse isotope effect because the dependence of T_c on the isotopic mass is opposite to what is expected from a straightforward application of the BCS (Bardeen-Cooper-Schrieffer) relation using a simplified form of the *McMillan* formula [5.113] written as

$$T_c \simeq \theta_D \exp(-1/\lambda_{ep}) , \tag{5.41}$$

where θ_D is the Debye temperature, and λ_{ep} the electron-phonon coupling parameter. Assuming λ_{ep} to be isotope independent, a decrease of the superconducting transition temperature T_c is expected in the sequence H, D, T due to the prefactor which is proportional to the inverse square root of mass. In binary metal hydrides consisting of a heavy metal and a light hydrogen atom, the electron-phonon parameter λ_{ep} can be written as the sum of an acoustical and an optical part due to the well separated spectra of the acoustical and optical phonons. The optical part of λ_{ep} is proportional to the electron density of states at the Fermi energy, the average of the square of the electron-phonon interaction matrix elements and inversely proportional to the product $m\langle\omega^2\rangle$, where m is the mass of the hydrogen atom and $\langle\omega^2\rangle$ the average of the square of the optical phonons. The product $m\langle\omega^2\rangle$ is called force constant. In an harmonic potential the force constants of H, D, T are equal.

Two models exist to explain the observed inverse isotope effect of the transition temperature in the Pd-X (X = H, D, T) system: in the model proposed by *Ganguly* [5.114] the inverse isotope effect is solely due to anharmonic effects. In the model proposed by *Miller* and *Satterthwaite* [5.115] different electronic structures in the vicinity of the hydrogen isotopes are considered to be the main cause of the inverse isotope effect. These changes are induced by the different zero-point vibrations because the root-mean-square displacement in the β-PdX phases is ≃0.23 Å for H and ≃0.20 Å for D at room temperature. The larger displacement in the case of the H atoms causes a greater overlap of the wave functions with those of the surrounding Pd atoms. This results in more electron states in the Pd-H bands than in the Pd-D bands. In this model the effects of anharmonic phonons on T_c are neglected.

Considerable theoretical effort has been devoted to these models. *Papaconstantopoulos* and coworkers [5.116] obtained very good agreement of their theoretical calculations (self-consistent band structure calculations using the augmented-plane-wave method with spin-independent relativistic corrections) with the experimental T_c curves using the *Allen* and *Dynes* [5.117] equation and the following ratio of the force constants

$$\frac{m_H \langle \omega_H{}^2 \rangle}{m_D \langle \omega_D{}^2 \rangle} = 1.2 \; . \tag{5.42}$$

The value of 1.2 was derived from older $PdD_{0.63}$ and $PdH_{0.63}$ neutron scattering data of *Rowe* et al. [5.118] and of *Rahman* et al. [5.119]. It is about 7% too large compared to the value of 1.12 which is an average of different experimental methods (bulk modules, thermal expansion, point-contact spectroscopy) [5.120]. Also the ratio of 1.12 was found in a recent neutron scattering experiment of PdH_r and PdD_r [5.30]. Therefore, a further effect is needed to explain the T_c curves quantitatively. *Griessen* and *de Groot* [5.119] obtained satisfactory agreement by the inclusion of Debye-Waller factors in the electron-phonon matrix elements of McMillan's theory. The contribution of the Debye-Waller factor to the inverse isotope effect stems from the mean square displacements of H, D, T.

Following the ideas of *Miller* and *Satterthwaite* [5.115], *Jena* et al. [5.121] showed recently that the electronic structure of PdH and PdD is influenced by the zero-point vibrations of H and D. They concluded that this quantum effect makes a small, but not negligible contribution to the T_c values and that the inverse isotope effect on T_c is dominated by the different force constants.

In conclusion, the inverse isotope effect on T_c cannot be explained by anharmonic effects alone. Other contributions due to the zero-point motion of the H, D, T atoms also have to be considered.

5.8 Isotope Dependence of Phonon Dispersion Curves in $PdD_{0.63}$ and $PdT_{0.7}$

Owing to their small mass, hydrogen atoms can easily follow the low energy vibrations of the surrounding lattice atoms. For long wavelength phonons, the hydrogen atom can move synchronously with the host atoms. In addition to these band modes, high energy vibrations of the hydrogen atoms are observed in a potential well provided by the lattice atoms (see also Sect.5.6). These oscillations are called optical modes because the hydrogen atoms vibrate collectively against the host metal atoms in stoichiometric hydrides. In very dilute metal-hydrogen alloys, the optical vibrations are called localized modes in contrast to the collective excitations. A third type of motion discussed in Sect.5.10 is diffusion.

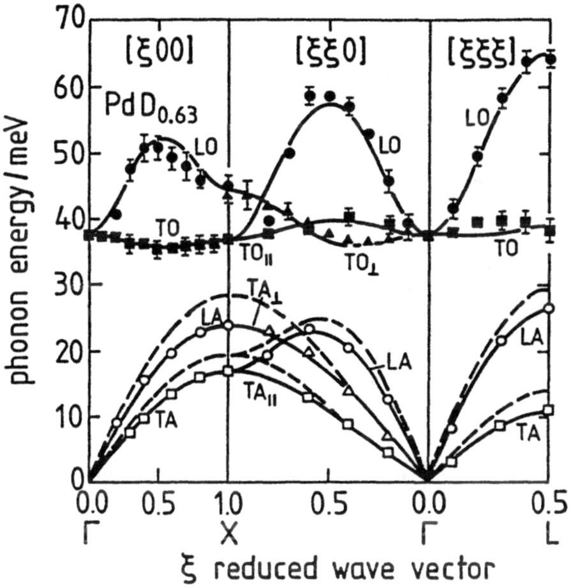

Fig.5.41. Phonon dispersion relation for $PdD_{0.63}$ at temperatures less than 150 K [5.118]

The phonons in a $PdD_{0.63}$ sample [5.118] are shown in Fig.5.41 as a function of the reduced wave vector for the principal symmetry directions. The results were obtained about 13 years ago for sample temperatures below 150 K. The dispersion of the acoustic phonons for pure Pd is included as dashed lines. The solid curves represent the fit by a Born-von Kármán model [5.118]. The energies of the phonons of the Pd lattice decrease with increasing deuterium concentration. There is only a marginal change in the shape of the acoustic dispersion curves. This effect is caused by a dilation of the lattice and a change of the electronic configuration by the uptake of hydrogen. It has also been seen in measurements of the Eliashberg function $\alpha^2 F(\omega)$, where $F(\omega)$ is the phonon density of state, and α^2 an energy dependent electron phonon matrix element, by means of point-contact spectroscopy [5.120]. The observed softening of the acoustic phonon branches at the boundaries of the Brillouin zone is the main cause for the lowering of the bulk modulus and the shear modulus C_{44} in β PdX_r (X= H, D, T) in comparison to pure Pd [5.38]. On the other hand, properties in pure Pd and β PdX_r which depend mainly on the transverse-acoustic phonons in the $\langle 110\rangle$ directions are expected to differ only slightly given the similarity of the $\langle 110\rangle$ TA branches at small wave vector. A quantity which depends sensitively on all the acoustic modes is the excess entropy of solution of hydrogen in metals [5.122].

In many other metal hydrides the opposite effect is observed. There occurs an increase of the energies of the acoustic modes with increasing hydrogen content [5.123].

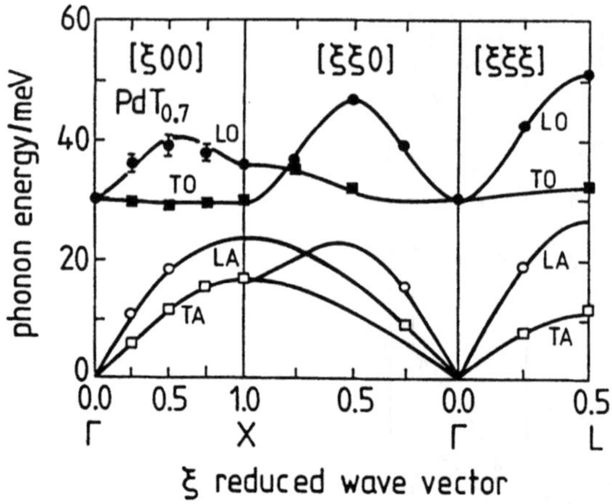

Fig.5.42. Phonon dispersion relation for $PdT_{0.7}$ at 80 K [5.124]. The reduced wave vector ξ is given in units of $2\pi/a$ with $a = 4.03 \cdot 10^{-10}$ m

The most interesting feature in Fig.5.41 is the large dispersion of the optical modes, especially of the longitudinal modes. This behaviour indicates strong second-neighbour D-D interactions. In contrast, the transverse-optical phonon branches are rather flat. To show again the variety of effects of hydrogen in metals it is mentioned that the optical modes of many other metal hydrides do not show any dispersion.

Figure 5.42 shows the recently published dispersion curves for $PdT_{0.7}$ [5.124] obtained at 80 K. A Pd single crystal was loaded with tritium. To avoid damage to the crystal during charging the T_2 pressure and sample temperature were increased to about 48 bar and 673 K. Under these experimental conditions it was hoped that the two-phase region $\alpha+\beta$ could be avoided [5.124]. This assumption is in contradiction to the critical parameters published in [5.37]. On the other hand, T_2 pressures of about 50 bar can only be produced in few laboratories. Also, the purity of the tritium gas used was excellent. The final tritium concentration in the Pd sample was $r = 0.71 \pm 0.03$ yielding a tritium activity of about 30 TBq ($\simeq$810Ci).

Table 5.7. Amplitude weighted energy moments of phonon dispersion relation in Pd-X (X = H, D, T) [5.124]

		$\langle E \rangle$ [meV]	$\langle E^{-1} \rangle^{-1}$ [meV]
$PdH_{0.6}$	Pd	16.5	14.4
	H	63.7	61.8
$PdD_{0.63}$	Pd	16.4	14.3
	D	43.3	41.7
$PdT_{0.7}$	Pd	16.1	14.0
	T	34.4	32.8

The differences between the acoustic modes of $PdT_{0.7}$ (Fig.5.42) and $PdD_{0.63}$ (Fig.5.41) are very small. The optical modes (which show about the same dispersion) are shifted strongly to lower energy values. They are still well separated from the top of the acoustic phonon branches. Table 5.7 gives a comparison of the amplitude weighted energy moments for the host lattice and the hydrogen isotopes calculated with the parameters of a Born-von Kármán model [5.124]. These values combined with the electron-phonon enhancement parameter for the acoustic and optical parts as calculated by *Papaconstantopoulos* et al. [5.116] and inserted into the superconducting transition temperature equation given by *Allen* and *Dynes* [5.117] do not allow the inverse isotope effect on the superconductivity in PdH and PdD to be reproduced. This result shows again (Sect.5.7) that anharmonicity alone cannot explain the inverse isotope effect in the Pd-X (X = H, D, T) systems quantitatively.

The very small differences between the acoustic branches of $PdD_{0.63}$ and $PdT_{0.7}$ indicate that the approximation made in Sect.5.1 of considering only the isotope dependence of the optical modes for phonon contributions was valid. On the other hand, the assumption that the phonons can be described with a harmonic model using only a single phonon energy is certainly poor.

5.9 Isotope Effect in Dingle Temperatures and Extremal Cross Sections of the Fermi Surface in PdH_r, PdD_r and PdT_r

In order to test the validity of the model proposed by *Miller* and *Satterthwaite* [5.115] several workers searched for isotope effects in the electronic properties of the PdH_r, PdD_r and PdT_r alloys.

De Haas-van Alphen experiments allow an accurate determination of extremal cross-sectional areas of the Fermi surface and of relaxation times for electron scattering (expressed in terms of the Dingle temperature T_D).

Such measurements were performed by *Bakker* et al. [5.125, 126] to study the isotope and concentration dependence of electronic properties in PdH_r and PdD_r. Recently, they extended their measurements to Pd loaded with tritium [5.127]. By quenching they avoided precipitation of the β phase in the PdX_r (X = H, D, T) samples with $r \leq 0.006$. Results for the concentration dependence of the Dingle temperature T_D and extremal cross sections A are listed in Table 5.8. A similarity in the isotope effects for the concentration derivatives of the extremal cross-sectional areas of several orbits and their Dingle temperature T_D is observed. At the points X, L and W of the Brillouin zone the cross-sectional areas which belong to the 4d hole Fermi surfaces decrease with increasing hydrogen concentration for every isotope, whereas the cross section of the large 5 s-p electron surface centred around the Γ point increases. An interesting, not yet understood isotope effect is observed for the hole pockets around X and L:

Table 5.8. Concentration derivatives of the extremal cross-sectional areas A of several orbits on special points of the Brillouin zone and of their Dingle temperatures T_D for PdH_r, PdD_r and PdT_r [5.125-127].

Fermi surface sheet	orbit centre	dlnA/dr (A = $\hbar$A/2πe)			dT_D/dr [K]		
		PdH_r	PdD_r	PdT_r	PdH_r	PdD_r	PdT_r
electron sheet	(0,0,0)	0.26(3)	0.24(3)	0.31(8)	750(159)	775(175)	690(180)
X hole pocket	X(0,0,1)	-2.0(2)	-3.7(4)	-2.4(3)	210(40)	445(59)	185(28)
X hole pocket	X(1,0,0)	-2.0(2)	-3.7(4)	-2.6(4)	225(25)	485(185)	194(48)
L hole pocket	L(½,½,½)	-8(1)	-14.6(2.0)	-10.8(3.0)			
hole orbit	W(½,0,1)	-1.3(2)	-1.7(3)		207(20)	226(25)	

$$\left.\frac{d\ln A}{dr}\right|_D \le \left.\frac{d\ln A}{dr}\right|_T < \left.\frac{d\ln A}{dr}\right|_H \tag{5.43}$$

and

$$\left.\frac{dT_D}{dr}\right|_T \le \left.\frac{dT_D}{dr}\right|_H < \left.\frac{dT_D}{dr}\right|_D . \tag{5.44}$$

At the point Γ and W no isotope effect is found within the experimental accuracy.

The experiments show that the observed isotope effects are confined to states on the hole ellipsoids at L and X. These represent only 3% of the total density of states at the Fermi energy E_F. The change in the 3% due to isotopic exchange is far too small to be observed in experiments such as those on the magnetic susceptibility, which measure the total density of states at E_F. This agrees with the susceptibility measurements of *Blaurock* [5.50] and *Wicke* [5.40, 128] where no isotope dependence for H and D in Pd was detected.

These results of *Bakker* et al. [5.125-127] indicate that in PdX_r (X = H, D, T) the isotopic dependence of electronic contributions at E_F is negligible and imply that ~1 electron per absorbed hydrogen atom in Pd fills states near E_F.

5.10 Diffusion of Protium, Deuterium and Tritium in Selected Metals

The motion of hydrogen atoms in metals by jumps between neighbouring interstices can be studied with macroscopic and microscopic methods. In the first category the change of a nonequilibrium spatial distribution towards equilibrium is observed, whereas in the second category motion is investigated under equilibrium conditions. The relaxation to equilibrium by hydrogen-transport over macroscopic distances can be studied with the Gorsky effect, resistivity, lattice parameter and permeation experiments, autoradiography, etc. The microscopic techniques include methods such as NMR, Mössbauer spectroscopy and quasi-elastic neutron scattering, where individual jumps are investigated.

Before discussing some diffusion experiments a short description of the diffusion processes themselves will be given.

In the classical theory [5.129, 130] the diffusion of the hydrogen atoms occurs via jumps over barriers. This behaviour is described by the equations:

$$D = D_0 e^{-E_a/kT} \tag{5.45a}$$

or

$$v = v_0 e^{-E_a/kT} \tag{5.45b}$$

where D is the diffusion coefficient, v the jump frequency, and v_0 the attempt frequency. E_a is the activation energy for diffusion and equals the height of the energy barrier between neighbouring interstitial sites. All hydrogen isotopes must overcome the same barrier. Thus in the classical theory E_a is isotope independent. On the other hand, the pre-exponential factors D_0 or v_0 are proportional to the inverse square root of the mass, and are thus isotope dependent.

A modified classical rate theory [5.131, 132] also yields an Arrhenius law similar to (5.45). Given their small mass, the hydrogen atoms are considered as localized oscillators with distinct energy levels. The activation energies are the differences between the minima of the potentials at the saddle point and at the interstice plus the differences in the vibrational energy levels. The prefactor shows the classical isotope dependence if the vibrational zero point energies are much lower than $k_B \cdot T$. Using these arguments, the isotope effect in fcc metals [5.133] was explained. For $\hbar\omega >> k_B T$, an isotope-independent prefactor [5.134] is found. This situation is observed in bcc metals.

Diffusion of hydrogen in metals is very often described by the small polaron model [5.135-138]. The hydrogen atom combined with the distortion field produced by the local expansion is termed a small polaron. In the following, only those aspects of the small polaron theory that are necessary for discussing the experimental results, are presented.

At low temperatures, diffusion occurs only via coincidence configurations between low-energy levels. The dominant jump process at very low temperatures is ground-state to ground-state tunnelling. The parameter E_a in (5.45) is proportional to the lattice activation energy necessary to establish a coincidence configuration. At these temperatures the activation energy for diffusion of the X (X = H, D, T) atoms is to a large part solely dependent on the properties of the metal lattice. The differences in the tunnelling matrix cause the isotope dependence of the prefactor.

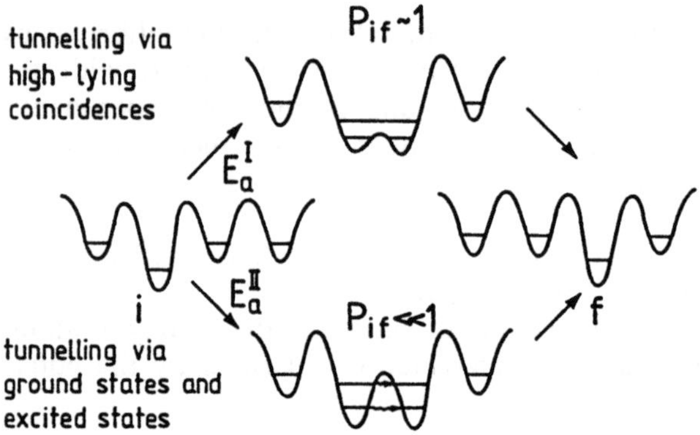

Fig.5.43. Schematic potential diagrams modelling the diffusion process of self-trapped hydrogen atoms from the initial (i) to the final (f) configuration. Only symmetric coincidence configurations are plotted for the adiabatic regime at high temperatures (top) and for the tunnelling via ground states or excited states at low temperatures (bottom)

At higher temperatures, phonon-assisted tunnelling processes become important. Coincidence configurations are generated by certain lattice modes. The host lattice atoms surrounding the self-trapped hydrogen atom are displaced and the vibrational energy levels are shifted and can coincide with energy levels in neighbouring unoccupied interstices. A path is now opened for a jump to a neighbouring site. Tunnelling processes via high-lying coincidence configurations in the adiabatic regime and via phonon-assisted formation of coincidence configurations at lower temperatures are shown schematically in Fig.5.43. Only symmetric coincidence configurations are plotted. In the adiabatic regime the H, D, T atoms respond adiabatically to the fluctuations of the surrounding metal atoms. These diffusion processes can also be described by an Arrhenius equation. The activation energy is only slightly isotope dependent, and the pre-exponential factor is determined by the Debye frequency of the host lattice.

The observation of a diffusion behaviour obeying (5.45) allows no decision as to which of the models discussed above is appropriate. At ordinary temperatures, classical theory, modified classical theory and small-polaron theory show more or less the same temperature dependence of the diffusion coefficient (5.45). Measurements with different hydrogen isotopes, however, and analysis of the isotope dependence of D_0 or v_0 and E_a, do allow discrimination between the different models.

5.10.1 Diffusion of H, D, T in V, Nb, Ta

The results of tritium diffusion measured by *Qi* et al. [5.74] with the Gorsky effect in a large temperature range are presented for V, Nb, Ta in Fig.5.44 and compared with H and D data [5.139-141]. In addition, the diffusion coefficient which was measured by *Matusiewciz* and *Birnbaum*

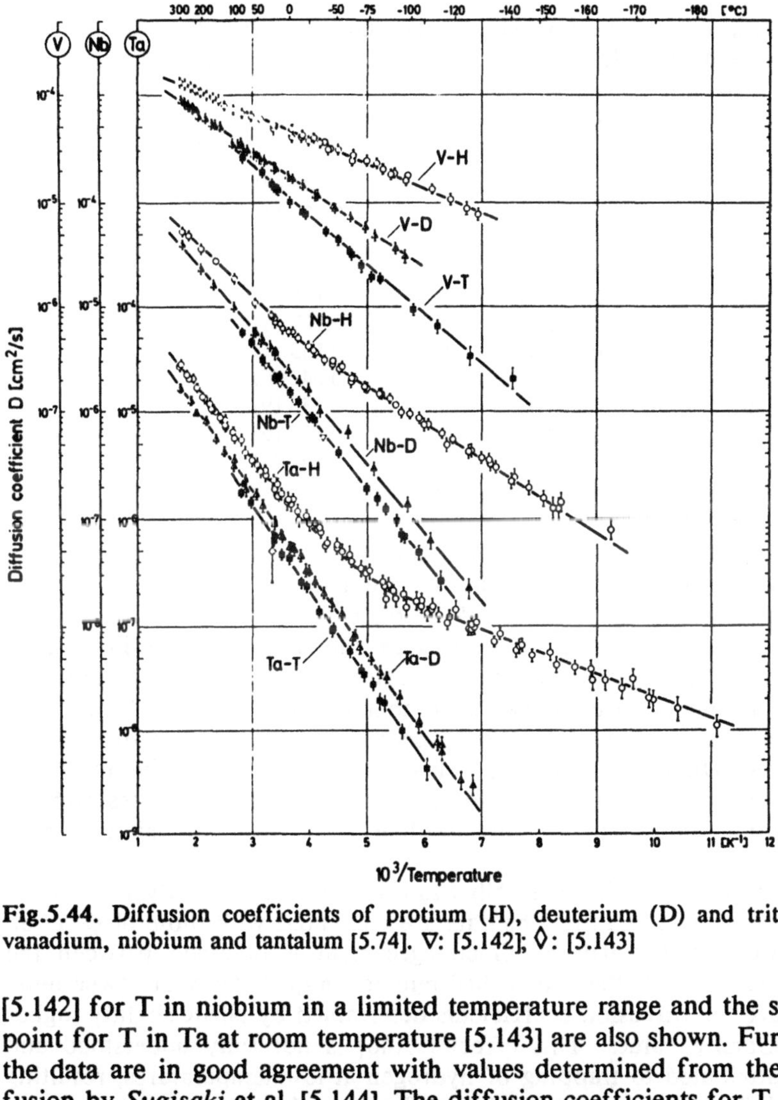

Fig.5.44. Diffusion coefficients of protium (H), deuterium (D) and tritium (T) in vanadium, niobium and tantalum [5.74]. ∇: [5.142]; ◊: [5.143]

[5.142] for T in niobium in a limited temperature range and the single data point for T in Ta at room temperature [5.143] are also shown. Furthermore, the data are in good agreement with values determined from thermal diffusion by *Sugisaki* et al. [5.144]. The diffusion coefficients for T in V, Nb, Ta [5.74] were measured at concentrations between 0.004 and 0.014. The data plotted in Fig.5.44 are extrapolated to infinite dilution.

At a given temperature the diffusion coefficient D for H, D and T in V, Nb and Ta decreases with increasing isotope mass obeying the inequality

$$D_H(T) > D_D(T) > D_T(T) \, . \tag{5.46}$$

The diffusion coefficients of the hydrogen isotopes differ by two orders of magnitude at lower temperatures. In the adiabatic regime, the isotope

Table 5.9. Parameters of the Arrhenius law $D = D_0 \exp(-E_a/kT)$ for diffusion of H, D, T in V, Nb and Ta [5.74]

	D_0 [10^{-4}cm^2/s]	E_a [eV]
V-H	3.1 (6)	0.045 (4)
V-D	3.8 (8)	0.073 (4)
V-T	5.6 (1.4)	0.094 (7)
NbH (T < 250K)	0.9 (2)	0.068 (4)
NbH (T > 250K)	5.0 (1.0)	0.106 (6)
Nb-D	5.2 (1.0)	0.127 (6)
Nb-T	4.4 (1.4)	0.133 (7)
Ta-H (T < 250K)	0.028 (15)	0.042 (6)
Ta-H (T > 250K)	4.2 (1.2)	0.136 (9)
Ta-D	3.8 (1.0)	0.153 (6)
Ta-T	3.7 (1.4)	0.162 (7)

effect is caused by an increase of the activation energies in the sequence, H, D, T. Within experimental limits the prefactors are equal. The D_0 and E_a values determined by fitting (5.45) to the experimental data are listed in Table 5.9. At about 250 K an interesting change of slope is seen in Fig.5.44 for Nb-H and Ta-H. The activation energies and prefactors are considerably lower at temperatures below 250 K than above 250 K. For D and T in Nb and Ta the same effect is also expected to occur, but at even lower temperatures. Diffusion constants below -100°C could not be measured because of the precipitation of hydrogen atoms in β-type phases and the long relaxation times. These would result in sensitivity and stability problems.

The diffusion of tritium in Nb was also measured recently with NMR by *Messer* et al. [5.145, 146]. The data are compared with those of *Qi* et al. [5.74] in Fig.5.45 for tritium. Good agreement is obtained between the values measured with these two different techniques. Only at lower temperatures are the diffusion coefficients determined by NMR slightly higher than the Gorsky values. This effect [5.146] is not fully understood and might be attributed to trapping of hydrogen at low temperatures, resulting in lower diffusivities than in a defect-free sample. The results presented above for bcc metals can nowadays be fairly well described by theory [5.147]. The situation worsens, however, for higher concentration MX_r (X = H,D,T) alloys for which interesting isotope effects have also been reported (see below). Improved theories require more accurate information on the interaction between hydrogen and metal atoms and on the anharmonic terms.

Bowman et al. [5.148, 149] reported large isotope effects in the activation energies for hydrogen jumps at higher concentrations in vanadium. An isotopic comparison of their data is only possible for the concentration r = 0.50 where the β phases are isomorphic below about 378 K (Sect.5.5.2). In

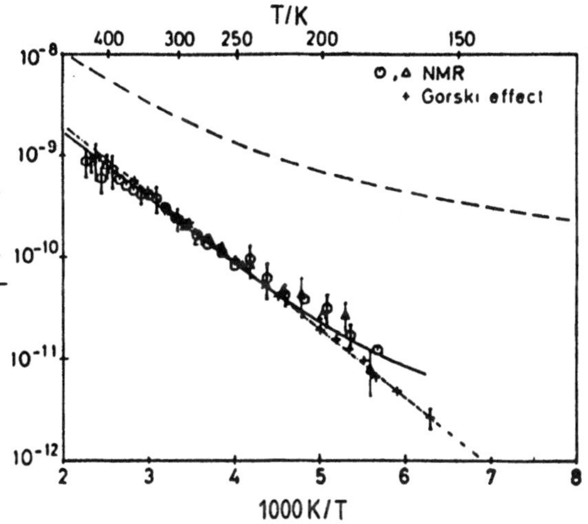

Fig.5.45. Comparison of the tritium diffusion coefficient in niobium [5.145]. +: Gorsky effect [5.74]; O,Δ: NMR [5.146]. Dashed line shows the proton diffusivity

the β-phase region, they obtained a higher diffusion coefficient and a lower activation energy for T than for H. These observations are the opposite of those at very low hydrogen concentrations (Fig.5.44 and Table 5.9). A complete description of this isotope effect (E_a^T = 0.31eV for T and E_a^H = 0.41eV for H in $VX_{0.5}$ (X = H,T) [5.148]) is still missing. *Bowman* et al. [5.148] explain the observed enhanced T mobility in $VT_{0.50}$ with atomic exchange processes between T atoms on nonequivalent interstitial sites.

5.10.2 Diffusion of H, D, T in Dilute Pd-Fe Alloys

In the high-temperature range, the diffusion coefficient of tritium in Pd [5.143] was determined again by *Sicking* and coworkers [5.26, 150]. The new value for the activation energy E_a (Table 5.10) fits into the sequence expected from the behaviour of H and D ($E_a^H > E_a^D > E_a^T$). At room temperature the diffusion of T in Pd is faster than that of D and H. This isotope effect is opposite to the one discussed in Sect.5.10.1 for low hydrogen concentrations in V, Nb and Ta. In addition, the diffusion coefficients of

Table 5.10. Parameters of the Arrhenius law $v = v_0 \exp(-E_a/kT)$ for the jump frequences of H, D, T in Pd (adiabatic region T > 100K) [5.150,154,155]

	v_0 [10^{12}s]	E_a [eV]
Pd-H	1.9	0.23
Pd-D	1.9	0.21
Pd-T	1.9	0.19

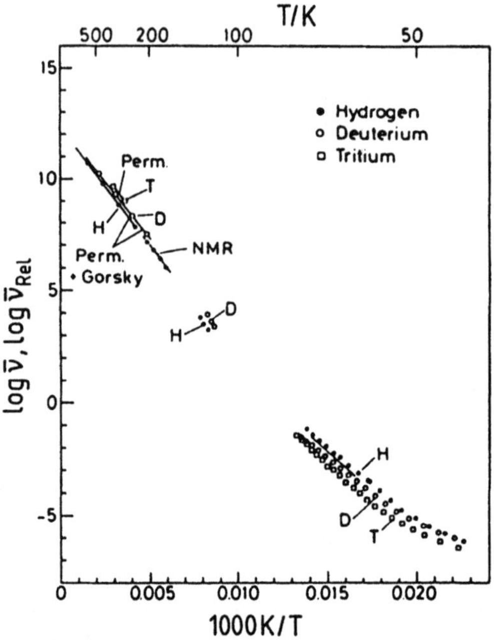

Fig.5.46. Jump frequencies ν_{rel} of protium, deuterium and tritium in $Pd_{0.95}Fe_{0.05}$ and ν in Pd [5.155]; full line between 50 and 60 K: [5.157]; NMR: [5.156]; Perm. + Gorsky: [5.157] and references therein. Perm.: [5.150]

H, D, T in Pd are lower than the corresponding values in V, Nb or Ta. This is a consequence of the larger activation energies. Thus tunnelling processes in fcc metals may not be detectable by techniques such as the Gorsky effect or permeation experiments. Further restrictions are imposed by precipitation of β-type phases.

Palladium has the lowest activation energies for diffusion of hydrogen atoms compared to all other fcc metals and is thus expected to be the most suitable fcc material in which to look for tunnelling processes. With the method of Magnetic After Effect (MAE), jump frequencies below 1 s^{-1} can easily be detected. Prerequisites for the MAE are a ferromagnetic material and hydrogen sites with a symmetry lower than the host lattice [5.151-153]. This can be achieved by alloying Pd with Fe. The trapping energy of H at substitutional Fe atoms in the Pd lattice was found to be below 10 meV [5.154]. Hence, the isotope effect measured for the reorientation of the Me-X (X = H, D, T) complexes in the Pd matrix should be representative of the X (X = H, D, T) diffusion in pure Pd. Figure 5.46 gives a summary of the jump frequencies of H, D, T in $Pd_{0.95}Fe_{0.05}$ and pure Pd obtained with different techniques [5.155].

Three temperature ranges may be distinguished with different characteristic diffusion processes [5.155].

Table 5.11. Parameters of the Arrhenius law $\nu = \nu_0 \exp(-E_a/kT)$ for the jump frequencies of H, D, T in Pd in the temperature range between 60 and 100 K [5.154, 155]

	ν_0 [10^7 s]	E_a [eV]
Pd-H	8.3	0.13
Pd-D	0.63	0.12
Pd-T	100	0.16

The adiabatic regime exists in the temperature range above about 100 K. Diffusion occurs via tunnelling processes between high lying coincidence configurations. The jump frequencies can be described by an Arrhenius law (5.45). The obtained preexponential factors ν_0 and the activation energies E_a are listed for the three isotopes in Table 5.10. As mentioned above, the activation energies decrease in the sequence H, D, T. This can be explained by different zero-point energies of the localized modes in the initial and saddle-point configurations due to the steeper potential in the saddle point. The pre-exponential factors are not isotope dependent, but are lower than the Debye frequency $\nu_D^{Pd} = 6 \cdot 10^{12}$ s^{-1}.

In the temperature range between 60 and 100 K, the experimental data can also be described by Arrhenius laws. The pre-exponential factors ν_0 and activation energies E_a of this temperature regime are listed in Table 5.11. Here, the activation energy for tritium is the highest of the hydrogen isotopes. This is in contrast to the isotope effect above 100 K. The (lattice) activation energy for tunnelling between ground states is approximately isotope independent. The experimentally observed differences in the activation energies can be explained by tunnelling processes between ground states and excited states. The transition temperature between the adiabatic regime and the regime below 100 K shifts to lower temperatures in the sequence H, D, T. Since in the case of T the vibrational energies are lower than for D or H, tunnelling processes via excited states become important for T at lower temperatures than for D or H.

In the temperature regime below 60 K, the jump frequencies are reduced even further. This regime may be attributed to the ground-state to ground-state tunnelling processes, although the tunnelling matrix elements become smaller. For a clear interpretation, the experiments need to be extended to lower temperatures.

5.10.3 Diffusion of H, D, T in Cu and Ni

Katz et al. [5.158] measured the diffusion coefficient of H, D, T in Cu and Ni in the temperature ranges from 720 to 1200 K and from 670 to 1270 K, respectively. Samples initially saturated with hydrogen were heated quickly to a certain temperature which was kept constant during the desorption

Table 5.12. Parameters of the Arrhenius law $D = D_0 \exp(-Ea/kT)$ for diffusion of H, D, T atoms in Cu and Ni

	Do [$10^{-3} cm^2/s$]	Ea [eV]	Temperature range [K]	Reference
Cu-H	11.31±0.40	0.403±0.003	720-1200	5.158
Cu-D	7.30±1.05	0.382±0.010	720-1200	5.158
Cu-T	6.12±0.51	0.378±0.006	720-1200	5.158
Ni-H	7.04±0.21	0.409±0.002	670-1270	5.158
	0.11	0.350	120-155	5.159
Ni-D	5.27±0.28	0.401±0.004	670-1280	5.158
	0.88	0.395	125-160	5.159
Ni-T	4.32±0.21	0.395±0.004	670-1270	5.158
	1.7	0.400	125-160	5.159

rate experiments. The values D_0 and E_a obtained from application of the Arrhenius relationship (5.45) are listed for H, D and T atoms in Table 5.12. *Katz* et al. [5.158] found that in the measured temperature range the diffusion coefficient follows the inequality: $D_T < D_D < D_H$. While the exponential prefactors D_0 show the approximate classical mass dependence ($D_0^H/D_0^D = \sqrt{2}$, $D_0^H/D_0^T = \sqrt{3}$) an inverse isotope dependence of the activation energies is observed which is in contradiction to classical theory ($E_a^H = E_a^D = E_a^T$). In the temperature range above about 700 K the activation energies for diffusion of H, D, T in Cu and Ni decrease with increasing hydrogen isotope mass. *Katz* et al. [5.158] had to consider anharmonic contributions to obtain reasonable values within the modified rate theory.

Hohler and *Schreyer* [5.159] performed magnetic after-effect measurements to determine the diffusion coefficient of H, D, T atoms in a number of dilute Ni alloys. They found that the jump frequencies did not depend on the sort of metal atoms alloyed with Ni and concluded that the data obtained are also true for pure Ni. The following isotopic dependence of the jump frequencies in the temperature range between 120 and 160 K was observed: $\nu_H >> \nu_T > \nu_D$. This means that tritium, although much less mobile than protium, diffuses faster than deuterium. If one uses again the Arrhenius equation (5.45) and the relationship between jump frequencies and diffusion coefficients, the values listed in Table 5.12 for the temperature range below 160 K are calculated. The inverse isotope effect of the activation energies can be explained qualitatively by a modified classical rate theory assuming higher vibrational frequencies in the saddlepoint than on the occupied interstitial sites. Especially for H in Ni distinction has to be made between a high- and a low-temperature range for diffusion, because the observed temperature dependence can not be described by a single Arrhenius equation. The same effect is not observed in the case of T in Ni, because, due to the larger mass of T, tunnelling processes become important at quite lower temperatures than for H. The observed diffusion behaviour $D_H >> D_T > D_D$ in special temperature ranges is a simple consequence of the different transition temperatures for the three hydrogen isotopes. A qualitative explanation of the discussed result is that at higher temperatures jumps over the barrier are dominant, whereas at lower temperatures tunnelling processes with smaller activation energies and prefactors become important.

The inverse isotope effect of the activation energies of H and D in Pd, Cu and Ni and the normal isotope effect in V, Nb, Ta (Sect.5.10.1) were explained by *Hauck* [5.160] with different crystal-field stabilization of H and D. With his theory he further claimed to be able to explain the normal and inverse isotope effects of H and D in many physical quantities, e.g. M-X (X = H, D) force constants, transition temperatures in superconductivity of M-X (X = H, D) systems, miscibility gaps, solubility, etc, and partly also their concentration dependence.

5.11 Lattice Dilation in V, Nb, and Ta Produced by Protium, Deuterium and Tritium

In a large variety of crystalline metals, alloys and metallic glasses, hydrogen absorption causes a volume expansion Δv which on the average is $2.9 \cdot 10^{-3}$ nm^3 per hydrogen atom [5.161]. The expansion is independent of hydrogen concentration (at least for low concentrations) and corresponds to a partial molar volume for hydrogen of 1.7 $cm^3 \cdot mol^{-1}$. Usually the volume change Δv per dissolved hydrogen atom is normalized by the atomic volume Ω of the host lattice atoms. With the simple relations that the volume V before hydrogen uptake is given by $V = N\Omega$, that the macroscopic volume increase ΔV due to the dissolved hydrogen atoms is given by $\Delta V = n\Delta v$, where N and n are the number of host and hydrogen atoms, and that $r = n/N$ one obtains the equation

$$\frac{\Delta V}{V} = r\frac{\Delta v}{\Omega} . \tag{5.47}$$

The left-hand side of the above equation can be determined experimentally by measurements of the volumes, of the length changes $\Delta L/L$ ($3\Delta L/L = \Delta V/V$ for small changes) and of the lattice parameter changes ($3\Delta a/a = \Delta V/V$). Length and lattice parameter changes can be measured with very high accuracy so that the precision of the actual $\Delta v/\Omega$ value is determined in most cases by the accuracy in the determination of the hydrogen concentration r.

A further method determines the density change that accompanies the dissolution of hydrogen in metals, using a microbalance for the measure-

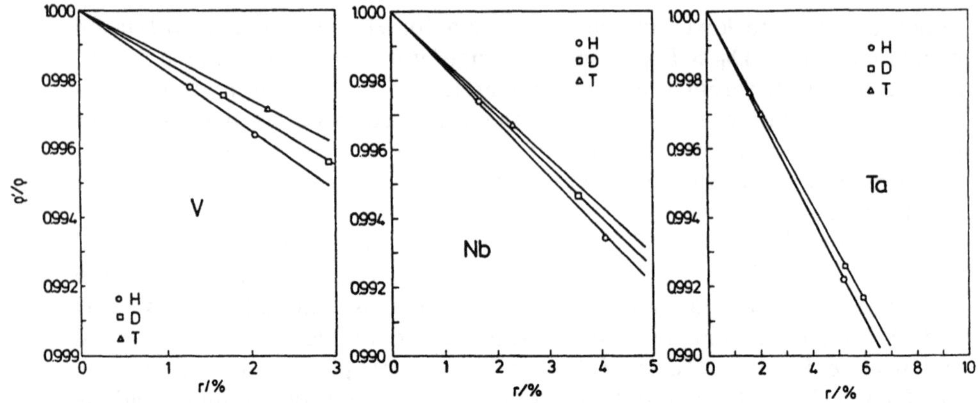

Fig.5.47. Density ratio ρ'/ρ of VX_r, NbX_r and TaX_r (X = H, D, T) samples as a function of concentration r [5.162]. ρ (ρ') is the density before (after) hydrogen isotope charging

ment of the sample weight in air and in a liquid. The $\Delta v/\Omega$ values can be calculated using the expression

$$\rho'/\rho = 1 - (\Delta v/\Omega - A'/A)r\ , \tag{5.48}$$

where ρ and ρ' are the densities of the metal before and after hydrogen charging. A′ and A are the relative atomic masses of the dissolved hydrogen isotopes and of the host atoms.

In the following, the results of accurate density measurements using the buoyancy technique, from work carried out on dilute hydrides, deuterides and tritides of V, Nb and Ta [5.162] are reported. The uncertainties in the density data are in the low 10^{-5} region. To improve on the accuracy, large pieces of V, Nb and Ta, weighing between 15 and 25 g, were used. Only in the case of the tritium-charged samples were samples of a few grams used due to the limited amounts of tritium available. The results are shown in Fig.5.47 and the values $\Delta v/\Omega$ are listed in Table 5.13. The purities of the H_2 and D_2 gas used for loading the samples were better than 99.7%. In the case of tritium, the atomic fractions of H, D and T in the gas were 0.02, 0.03 and 0.95, respectively. Figure 5.47 shows that the density

Table 5.13. $\Delta v/\Omega$ values for protium (H), deuterium (D) and tritium (T) in vanadium, niobium and tantalum as determined with precision density measurements [5.162]

	H	D	T
V	0.196 (3)	0.191 (3)	0.188 (4)
Nb	0.172 (1)	0.1695 (10)	0.175 (3)
Ta	0.154 (1)	0.152 (1)	0.156 (3)

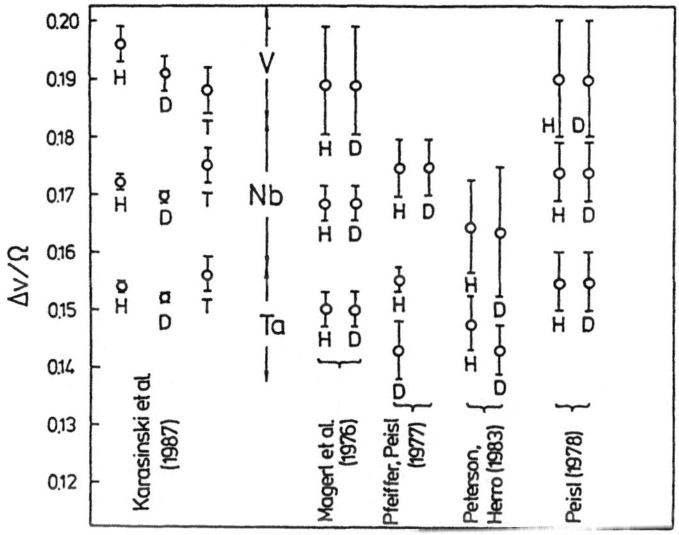

Fig.5.48. Comparison of relative volume change $\Delta v/\Omega$ determined with precision density measurements [5.162] with results of previous work [5.162-165]

of the three metals decreases linearly with the concentration r, as expected from (5.48). The listed data for $\Delta v/\Omega$ in Table 5.13 do not show an isotope effect for Nb and Ta. For Ta, this result is in disagreement with the data of *Pfeiffer* and *Peisl* [5.163], who reported that protium expands the lattice about 8% more than deuterium. A very small isotope dependence of the partial molar volume of the H, D and T atoms is observed for vanadium, such that the volume expansion decreases slightly with increasing hydrogen isotope mass. A comparison of $\Delta v/\Omega$ values obtained by different groups [5.162-165] is given in Fig.5.48.

6. Properties of Helium-3 in Selected Metals

An understanding of the behaviour of He in metals is important because He is generated in charged tritium storage media and in the structural materials of fission reactors, accelerators, breeders and future fusion reactors. Helium accumulation causes deterioration of the material properties and may influence the useful lifetime of reactor components.

For the study of these effects, metals can be charged with He in nuclear reactors, by ion implantation or by using the "tritium trick" [6.1] (Chap. 1), but only in few cases from the gas phase. The first two methods introduce lattice damage into the metal, the only exception being subthreshold implantation. The interaction of He with the lattice defects created during implantation influences the behaviour of He drastically. A disadvantage of the subthreshold implantation method is that only a thin layer of about 10 nm at the surface is doped with He. Since the threshold energies for atomic displacements lie in the range 10 to 40 eV [6.2], only low He-ion energies of up to about 50 and 400 eV for light and heavy metals, respectively, can be used. Furthermore, when using subthreshold ion implantation for heavy metals it is very important to avoid surface contamination with low-Z atoms such as oxygen, carbon, sulphur, etc. By (n,α) processes large samples can be doped in reactors. However, the whole sample will be activated.

In contrast, the "tritium trick" produces no lattice damage in the metal since the average (maximum) recoil energy of about 0.4 (3.4) eV transferred to the ^{3}He atom during the T decay is far too low to create Frenkel defects. Large samples can be charged homogeneously with T and, therefore, with ^{3}He. Using only a single metal tritium sample, it is often possible to study the physical quantities of interest as a function of He concentration since the latter increases with age. To calculate the He concentration with (1.2), information on the tritium distribution is required since the tritium uptake - measured by pressure drop or by weight increase - yields only an averaged concentration. In two-phase situations or when trapping at dislocations occurs, different tritium concentration may coexist in the sample. Disadvantages of the "tritium trick" are 1) that the method only works well for materials which absorb hydrogen easily, 2) that the aging times to obtain certain He concentrations can be very long due to the tritium half-life of 12.323 years (e.g., after one year (month), 5.5% (0.46%) of the initial tritium will be converted into ^{3}He) and low initial tritium concentrations, 3) that an interaction between the tritium and He atoms may exist, and 4) that highly radioactive samples have to be handled. In connection with the last statement it should be remembered (Sect.2.2) that external exposure to T gas is not hazardous.

A rarely used method to charge metals with He is the application of very high pressures. In the case of Au [6.3] at He gas pressures up to 3 kbar, a linear relationship between He concentration and pressure was obtained. In the case of Ni, such high pressure experiments did not give conclusive data on the bulk solubility [6.4,5]. Before discussing experiments, a few well-known properties of He in metals will be presented.

6.1 Basic Properties of Helium in Metals

6.1.1 Energetics of Helium in Metals

For a full understanding of the changes of macroscopic mechanical properties due to He, the microscopic quantities such as formation energy of interstitial (substitutional) He, E_I^f (E_S^f), migration energy of interstitial He E_I^m, binding energy of two interstitial He atoms at one interstice, E_{2I}^b, binding energies of He vacancy, E_V^b, He grain-boundary E_{GB}^b, and He-dislocation complexes E_D^b, activation energies for diffusion E_a, etc. should all be known. The meaning of some of the quantities mentioned is shown schematically in Fig.6.1. Examples of the relations between these quantities are: The formation energy for substitutional helium, E_S^f, is the sum of the formation energy of an interstitially dissolved helium atom, E_I^f, and of the vacancy formation energy E_V^f, minus the He-vacancy binding energy E_V^b. The dissociation energy E_V^d of a He atom from a vacancy is the sum of the He-vacancy binding energy E_V^b and of the interstitial migration energy E_I^m. In addition the following equation holds: $E_V^d = E_I^f + E_I^m - E_S^f + E_V^f$. On the basis of these energies, the equilibrium between the gas phase and the He atoms dissolved in a metal can be discussed in analogy to the solubility of

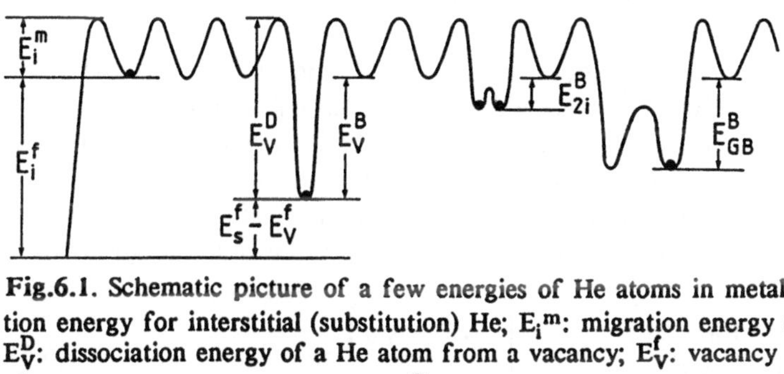

Fig.6.1. Schematic picture of a few energies of He atoms in metals, E_i^f (E_s^f): formation energy for interstitial (substitution) He; E_i^m: migration energy for interstitial He; E_V^D: dissociation energy of a He atom from a vacancy; E_V^f: vacancy formation energy; E_V^B: He-vacancy binding energy; E_{2i}^B: binding energy of two He atoms on one interstice; E_{GB}^B: He-grain boundary binding energy

hydrogen in metals (Sect.5.1). The situation here is somewhat simpler since the He gas atoms have only translational energy. In the regime of the ideal gas law, the following equations describe the chemical potential of one He atom in the gas phase

$$\mu_{He}^{gas} = kT\cdot\ln\left(\frac{p}{kT}\frac{h^3}{(2\pi m\ kT)^{3/2}}\right) \tag{6.1}$$

and in the metal on substitutional (S) sites

$$\mu_{HeV}^{metal} = H_S^f - TS_S^f + kT\cdot\ln\left(\frac{c_S}{1 - c_S}\right) \tag{6.2}$$

or on interstitial (I) sites

$$\mu_{HeI}^{metal} = H_I^f - TS_I^f + kT\cdot\ln\left(\frac{c_I}{N - c_I}\right) . \tag{6.3}$$

Here c_S and c_I are the ratios of He atoms on substitutional and interstitial sites to the metal atoms, k is the Boltzmann constant, h is the Planck constant, m the mass of a He atom, and N gives the number of interstitial sites per metal atom. In the following, the term pV is neglected yielding $E_I^f \simeq H_I^f$ and $E_S^f \simeq H_S^f$.

Thermodynamic equilibrium yields the following equations

$$c_S = \frac{h^3}{(2\pi mkT)^{3/2}}\frac{p}{kT}\exp\left(-\frac{E_S^f - TS_S^f}{kT}\right) \quad \text{and} \tag{6.4}$$

$$\frac{c_I}{N - c_I} = \frac{h^3}{(2\pi mkT)^{3/2}}\frac{p}{kT}\exp\left(-\frac{E_I^f - TS_I^f}{kT}\right) . \tag{6.5}$$

In the case of Ni, concentrations $c_S \simeq 6\cdot10^{-9}$ and $c_I/N \simeq 1\cdot10^{-13}$ are obtained using the values $E_S^f = 3.1$ eV and $E_I^f = 4.5$ eV given in [6.6], $S_S^f = S_I^f \simeq 8$ k, $p = 10^6$ bar and $T = 1500$ K. For substitutional occupation, the higher concentration value is caused by the lower substitutional formation energy due, in turn, to the high binding energies E_V^b of He atoms to vacancies (for $E_V^b > E_V^f$). As a consequence, c_S will always be much larger than c_I

$$\frac{c_S}{c_I} = N\exp\left(-\frac{E_S^f - E_I^f}{kT}\right) = N\exp\left(+\frac{E_V^b - E_V^f}{kT}\right) >> 1 . \tag{6.6}$$

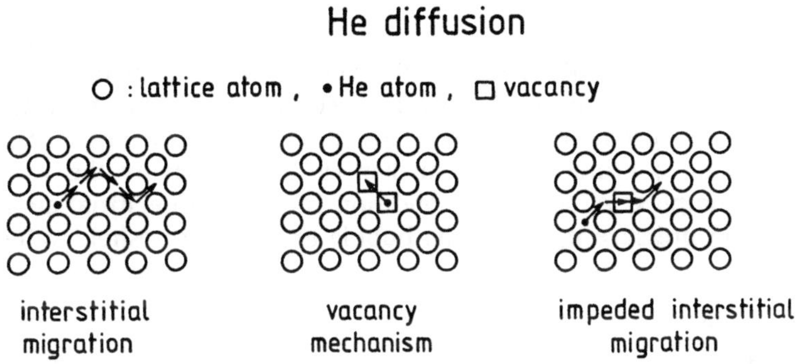

Fig.6.2. Schematic representation of three possible diffusion mechanisms in a lattice [6.6]

At very high pressures, He deviates from ideal gas behaviour (assumed above) and the concentrations c_S and c_I may increase by even 3 to 4 orders of magnitude [6.7]

These extremely low solubilities of He are the reason for cluster or bubble formation even at small concentrations.

6.1.2 Kinetics of Helium Precipitation in Metals

Helium bubble formation requires diffusion of the He atoms. There are three different models [6.6] for the migration of He in metals:

1) Migration via jumps between interstitial sites. These processes should be similar to the diffusion discussed in Sect.5.10 for T in metals. On the left-hand side of Fig.6.2 the diffusion mechanism by interstitial migration is shown. Due to the large binding energy between He atoms and other defects, this type of diffusion can only occur in defect-free regions.
2) Migration via the vacancy mechanism (see centre of Fig.6.2). The substitutional He atom changes place with a vacancy in its immediate neighbourhood.
3) Migration of He via the dissociation from a He-V complex and subsequent interstitial diffusion. The activation energy for this He diffusion process is $E_a = E_V^d - E_V^f$. This diffusion mechanism is shown on the right-hand side of Fig.6.2.

In most simulation experiments both vacancies and He atoms are generated. Both migrating particles contribute to bubble nucleation and growth.

In contrast, He bubbles in T-loaded samples (see Sect.6.3) aged at low temperatures (below $T_m/3$ where T_m is the melting point) are formed without vacancies. In such materials most of the He atoms are generated in interstitial positions. He atoms migrate with the low migration energy discussed in process 1 until they cluster or are trapped at impurities or defects

[6.8,9]. If the number of He atoms in the clusters exceeds 5-7, then according to theoretical calculations [6.8,9], energy can be gained by the spontaneous formation of a Frenkel pair. The vacancy is occupied by He atoms (self-trapping mechanism [6.9]). This process of Self-Interstitial Atom (SIA) emission continues under further He absorption by the vacancy-He cluster-complex. A necessary condition for metal interstitial atom emission is that the pressure p_I inside the bubble exceeds a threshold value [6.7] given by

$$p_{SIA} \simeq \frac{2\gamma}{\ell} + \frac{E^f_{SIA}}{\Omega} , \tag{6.7}$$

where γ is the surface tension of the metal, E^f_{SIA} the formation energy of P_{SIA}, ℓ the radius of the bubble, and Ω the atomic volume. P_{SIA} may reach values of up to 10^6 bar.

After reaching a certain bubble size, the punching out of dislocation loops is energetically more favourable. The threshold pressure for loop punching has been estimated [6.7] to be

$$p = (2\gamma + \bar{\mu}b)/\ell \tag{6.8}$$

where $\bar{\mu}$ is the isotopic average of the shear modulus, and b the Burgers vector of the dislocation loop.

Growth of He bubbles without gas supply may occur either by migration and coalescence or by Ostwald ripening [6.7]. Coalescence means that the centre of gravity of the bubbles moves, due to diffusion of host metal atoms in the bubble, from one site to the other. He atoms and He bubbles along the path of the bubble are absorbed. In Ostwald ripening, large He bubbles grow at the expense of small ones since there is a concentration gradient between bubbles of different size. In the neighbourhood of small bubbles with large He pressure, a higher concentration of dissolved He atoms exists than in the neighbourhood of large bubbles. Diffusion down the concentration gradient leads to an increase in the size of the larger bubbles.

Further discussion of the above phenomena with regard to kinetics, temperature, He and/or vacancy concentration, microstructural effects and/or interaction with other defects and defect fluxes can be found elsewhere [6.6,7,10-16].

6.2 Helium Release from Metal Tritides

Helium release behaviour has been extensively studied for metals such as Sc, Ti, Zr, Er and U, which can form very stable tritides with high concentrations (Sect.3.6) and have, therefore, high He production rates. The main result is that during the first years of aging, the He release rate is less than 2% of the generation rate for many ditritides [6.17-19]. When the atomic ^3He concentration reaches critical values of about 0.1 to 0.3 (depending on the metal) the He release rate at constant temperature increases drastically and reaches, or even exceeds, the generation rate. The critical concentrations are about 0.3 for $TiT_{2-c}He_c$ [6.19], 0.25 for $ErT_{2-c}He_c$ [6.19] and 0.11 for $UT_{3-c}He_c$ [6.18,20]. "Accelerated release" occurs if the ^3He release rate roughly equals the ^3He generation rate. The latter was approached in Sc and U tritides after about 900 [6.19] and 1000 [6.18] days, respectively. The release rates were also found to be particle-size dependent. UT_3 powder, with an initial surface area of 0.6 m^2/g, released 4% of the produced He, whereas UT_3, with an initial surface area of 3.0 m^2/g, released 21% of the He after 650 days [6.19]. In the case of a thin $ScT_{0.3}$ film, the release rate was about 5 times larger than for a single crystal [6.19]. The temperature dependence of helium release from erbium tritide films was investigated by *Mitchell* and *Patrick* [6.21]. After a temperature rise they observed a large increase of the release rate which then dropped slowly to a constant level slightly above the value at the lower temperature. A strong temperature dependence of the critical He concentration was found [6.21]. Furthermore, ErT_2 samples displaying accelerated release could be brought back into a state of low He release by high-temperature cycles [6.21]. *Mitchell* and *Provo* [6.22] reported a large scatter in the He release rate from ErT_2 and ZrT_2. It was at the highest level during the transition into the accelerated release. The variations in the release rate could be also stimulated by vibrating and flexing the films. The explanation for the large bursts of about 10^9 He atoms is that He is released in a cascading process where parts of the He bubbles lose their He inventory. Even young ErT_2 samples could be stimulated by vibrations to a collective release, although their He release rate was below 1% [6.22]. The critical ^3He concentrations given above could be calculated by *Spulak* [6.23] using recent observations of the He behaviour.

Thomas et al. [6.23,24] performed He release measurements for stainless steel, Al and Ni samples and found the following results: 1) He is mobile above 100 K; 2) no release occurred during 80 K aging; 3) He is deeply trapped, even in single crystals of Ni; 4) the released He fraction was dependent on the initial He concentration suggesting a He-He interaction; 5) only a small fraction of the total ^3He content was released at temperatures as high as 1000 K in Ni and stainless steel. Similar studies employing isochronal and isothermal anneals for various metals were made by *Cost* and *Hickmann* [6.25]. Isochronal anneals of $UT_{3-c}He_c$ samples aged for different periods [6.26] showed two release peaks, one due to desorbing T_2 gas at about 200°C and the other due to ^3He atoms, which was shifted strongly to lower temperatures with age. A recent paper [6.27] discussing the ^3He retention behaviour in $VT_{0.50-c}He_c$ and $VT_{0.75-c}He_c$ reports that most of the ^3He generated is trapped in microscopic bubbles during the observation time of 2200 days. $PdT_{0.6-c}He_c$ has been found [6.26] to retain essentially all ^3He generated during 3.6 years.

The problem of He release is of importance in relation to the use of metals as tritium storage materials. An extrapolation of the behaviour reported above shows that after a few half-lives of tritium about 70% to 90% of the decay product He will be in the gas phase. No significant release of tritium due to ^{3}He-induced effects has ever been reported. LiT is an exception; here T_2 molecules produced by radiolysis were found in the He-filled bubbles with pulsed NMR techniques [6.26, 28]. Such tritium may be discharged to the outside after bubble rupture. If one takes into account the pressure increase due to the released He in MT_r storage containers, metal tritides with low equilibrium pressures can be considered as safe tritium storage materials. A further conclusion is that the oxide layers are far more permeable to He than to tritium (Sect.4.2).

6.3 He Bubbles in Metal Tritides

During irradiation of a metal, helium may become trapped by radiation-induced lattice damage. Clustering of He produced by He-ion implantation with energies below the displacement threshold energy or by the tritium trick was only recently established. Important steps for the understanding were the development of the self-trapping model [6.9] (Sect.6.1.2) and the direct TEM observation of He-bubbles in metals without radiation-induced lattice displacements. Before discussing these results, the work of some other groups from which it was more indirectly concluded that He in T-charged samples precipitates in bubbles, should be mentioned.

6.3.1 NMR Measurements

The distribution of ^{3}He in LiT, TiT_2 and UT_3 was investigated by *Bowman* and *Attalla* [6.20, 29] using pulsed NMR techniques. They explained the NMR line shapes and nuclear relaxation times by assuming that most ^{3}He atoms are retained in microscopic gas bubbles with dimensions smaller than 0.01 μm. In recent NMR measurements [6.30], microscopic bubbles were also observed in ternary alloys such as $ZrNiT_r$ and Mg_2NiT_r. This behaviour, which is also found in ionic, transition-metal and actinide tritides [6.31], is believed to occur in most materials and can be explained by the self-trapping model [6.9].

6.3.2 TEM Measurements

The hypothesis of bubble growth by loop punching is supported by direct experimental observations: dislocation loops and He bubbles were found after subthreshold He implantation in Mo [6.32] and Au [6.33]. An interpretation of both experiments by means of the self-trapping model [6.8, 9] is hampered by the fact that in the case of Mo nucleation sites were formed using a pre-damage bombardment. In the case of Au a thermal treatment of the sample was necessary.

Bubbles were directly observed with TEM in the absence of lattice defects and temperature treatments in $PdT_{0.6}$ [6.34], VT_r [6.35-37], $NbT_{0.01}$ [6.24] and $ZrT_{1.6}$ [6.38, 39].

TEM Measurements of $PdT_{r_0-c}He_c$ Samples

Using an equilibrium pressure of about 0.5 bar T_2, $PdT_{0.6}$ [6.34] samples were aged at room temperature to a He concention of 6000 ppm. They were subsequently thinned for TEM analysis. He bubbles and dislocation loops were readily visible after only 66 days of aging.

TEM Measurements of $VT_{r_0-c}He_c$ Samples

VT_r samples charged to the two-phase region (Sect.5.5.2) were studied with TEM as a function of time for 14 months [6.35-37]. This allowed a simultaneous study of the helium behaviour in α and β phase regions. The latter had about 10 times higher He generation rates than the α phase (Sect.5.5.2). The main results were: after about 7 to 14 days of aging, a rather dense distribution of small interstitial dislocation loops was observed in the β domains of the V-T alloy [6.35]. With increased age, the dislocation loop structure became coarser and the diameters of the loops larger, resulting finally in a dislocation network. After aging times of about 13 weeks, small ^3He bubbles with a density of $3.5 \cdot 10^{23}$ m^{-3} and a mean diameter of 1.2 nm were observed in the β phase. In the α phase VT_r areas, bubbles and tube-shaped cavities were found along dislocations and at low-angle grain boundaries. Isolated ^3He bubbles were observed which had relieved their high internal pressure by prismatic loop punching, see (6.8) and Fig.6.3. The distribution of He bubbles was homogeneous in the β phase, whereas He bubbles appear to form heterogeneously in the α phase. Because of the 10 times lower ^3He generation rate in the α phase, the role of pre-existing defects is more important in this phase. Such defects may even become the dominant sites for helium clustering and bubble growth [6.37].

TEM Measurements of $NbT_{r_0-c}He_c$ Samples

This effect was studied by *Thomas* [6.24] using NbT_{r_0} samples with low initial tritium concentrations of $\simeq 0.01$. No bubbles at all were found in large parts of the samples aged at room temperature to an average of 3000 ppm He. Only dislocations were decorated with ^3He bubbles. A rather high density of dislocations was observed, although the sample had been annealed to an average grain size of $\simeq 1$ mm diameter before charging. As might be expected [6.40] the decoration of dislocation lines by strings of

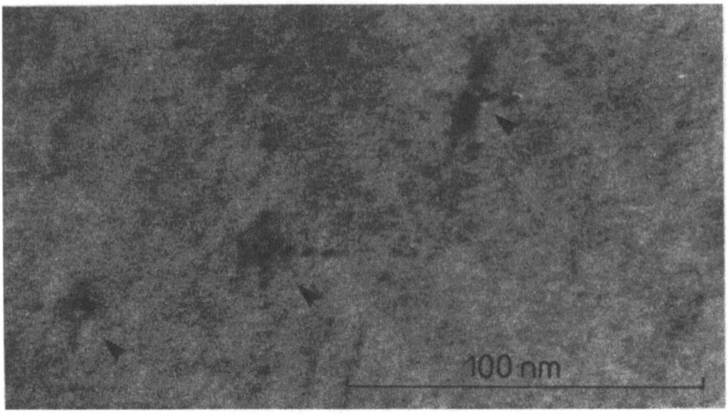

Fig.6.3. TEM micrograph [6.35-37]: Bright field. Arrows denote ^{3}He bubbles in α phase areas of a VT_r sample aged for 80 days. These bubbles have relieved their large internal pressure by punching out dislocation loops

bubbles or even tubes filled with ^{3}He has a dramatic effect on the mechanical properties. Even low-level stress resulted in intergranular fracture [6.24]. Two models have been proposed to explain the preferential ^{3}He decoration of dislocation lines [6.24, 37]: 1) He generated at low rates from low tritium concentrations is simply trapped on dislocations, 2) the initial tritium concentration is inhomogeneous with the formation of concented Cottrell clouds along dislocations [6.42] which result in higher He production rates. Again, these He atoms will be trapped by the dislocations.

TEM Measurements of $ZrT_{r_0-c}He_c$ Samples

A long-term TEM study of ZrT_r samples [6.39] (for preliminary results, see [6.38]) may be summarized as follows: 1) He bubbles of $\simeq$1 nm diameter were observed after ~3 weeks of aging. 2) Loops, network dislocations and phase boundaries were strongly decorated with He bubbles. 3) After about 50 days the apparent density of isolated He bubbles was time independent. A value of about $5\cdot10^{23}$ m^{-3} was determined. This indicates that bubble nucleation must have taken place at a very early stage. 4) The bubble size distribution shifts slightly towards larger diameters with age. This effect is shown in Fig.6.4. The mean bubble diameter increases approximately as the cube root of the aging time indicating constant He density in the bubbles and no He release. 5) Most bubbles showed strain contrast indicating internal pressures above 20 kbar. 6) After aging for 838 days at room temperature, facetted bubbles could be clearly observed indicating a self-diffusion process. Surface diffusion constants of the order of 10^{-26} m^2/s were deduced [6.39]. 7) The network dislocation density ρ was estimated to stay in the range $10^{15} < \rho < 10^{16}$ $[m^{-2}]$ during aging. The apparent constancy of the network dislocation density can be explained by the balance of the generation and annihilation of dislocations via adsorp-

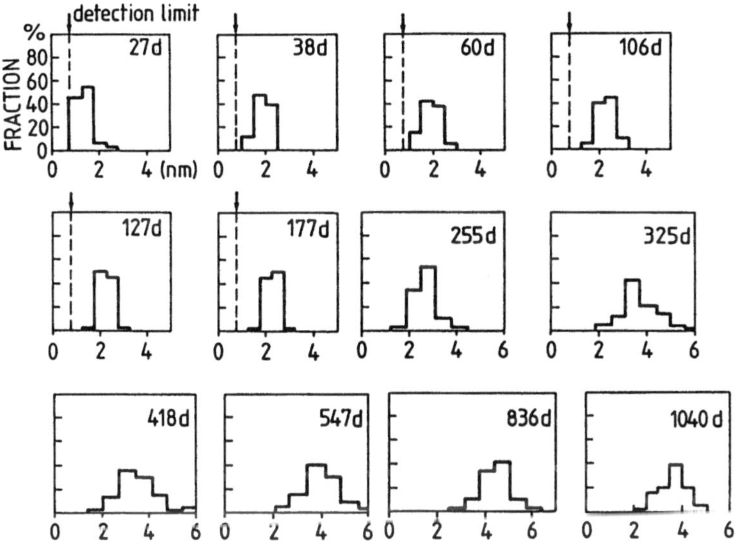

Fig.6.4. Distribution of the observed diameters of ^{3}He bubbles as a function of aging time [6.38, 39]

tion of self-interstitial atoms emitted by bubble formation. 8) A second generation of tiny bubbles formed at dislocations with bubble diameters between 2 to 3 nm. The concentration of He trapped in these bubbles in δ-$ZrT_{1.6}$ was estimated to be about 1% after about 836 days. The corresponding value in the bubbles of the first generation was about 9%. The calculated He concentration using (1.2) was about 19%. The discrepancy is ascribed to uncertainties in measuring the bubble density distributions, the volumes, the number of He atoms per bubble and the release rate to the outside. 9) In samples aged for more than ~2 years, an interconnected system of He-filled channels was observed, in addition to the dislocation network. Figure 6.5 shows a TEM micrograph of $ZrT_{1.6}$ aged for 4 years. ^{3}He bubbles and a channel-like network of dislocations decorated with ^{3}He bubbles are easily visible. A schematic view of the dislocations and bubbles is shown in Fig.6.6. This feature may explain the observation of bursts of at least 10^9 He atoms (Sect.6.2) occurring in the state of accelerated release in certain tritides [6.22]. A sudden connection of one of the tubes to an external surface may release a certain amount of ^{3}He to the outside leading to a pressure drop in the interconnected channels. Conceivably, this pressure drop may cause the channels to close. Re-opening of the channels is only possible after refilling with He atoms. 10) No significant difference in the aging behaviour of the γ and δ phases in Zr was detected, although the concention ratio of the two phases is 1.6.

Thus, three different states of aging may be distinguished: 1) the generation of the bubbles within the first 50 days; 2) growth of the bubbles, formation of a dislocation network and decoration of the dislocations, 3)

Fig.6.5. TEM micrograph of a δ-$ZrT_{1.6}$ area aged for approximately 4 years [6.39]. Visible are ^3He-filled bubbles and an interconnected system of ^3He-filled channels

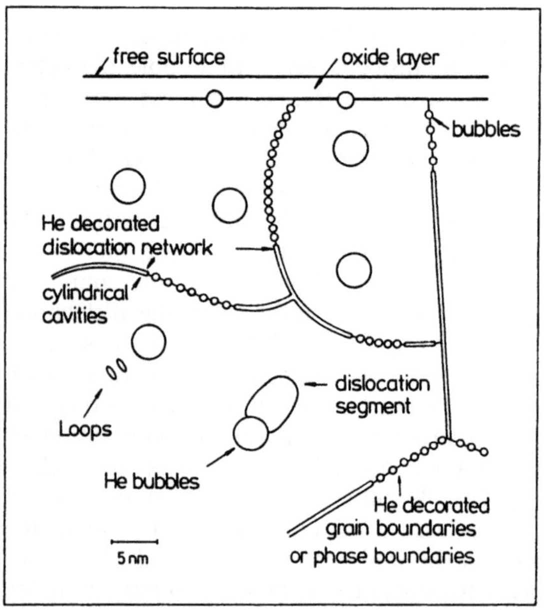

Fig.6.6. Schematic diagram of the microstructural elements occurring in δ-$ZrT_{1.6}$ after an age of about 3 years [6.39]

the development of interconnecting ^{3}He channels giving rise to the accelerated release. It is believed that the behaviour reported above for ZrT_x is typical for most metals.

6.3.3 Observation of Solid-Fluid Transitions of ^{3}He in the Bubbles of $PdT_{r_0-c}He_c$ Samples

A solid-fluid transition of ^{3}He bubbles [6.41] was observed near 250 K in $PdT_{0.586}He_{0.034}$ material aged for one year. The results obtained from NMR relaxation time measurements show an abrupt change of the spin-lattice relaxation time near 220 K as strong evidence for a phase transition and a coexistence region of the two phases between 220 and 280 K, which is probably caused by the ^{3}He density distribution in the bubbles. The ratio of ^{3}He to host-atom vacancies is found to be 2.0±0.2, a value which is in fair agreement with the conclusions of swelling results (Sect.6.5). In addition, the activation energy for ^{3}He diffusion in the solid phase under a pressure of about 60 to 110 kbar was determined to be 0.032 eV.

6.4 Acoustic Emission by Metal Tritides

Precipitation in dilute metal hydrogen alloys may lead to Acoustic Emission (AE) [6.43,44].

Schober et al. [6.45] observed sound waves emitted by metal tritides with a conventional lead zirconate titanate transducer [6.46]. After a short incubation period the samples discontinuously emit sounds waves. A typical diagram of the averaged acoustic emission as a function of aging time is shown in Fig.6.7, where the acoustic emission rate is integrated over a time of 5 h. The AE is believed [6.45] to be caused by collective emission processes of dislocation loops, since the emission from a single dislocation loop cannot be detected by AE methods. The collective loop punching processes are only observed in a certain intermediate state of the microstructural evolution, mainly after the formation of very small bubbles accompanied by closely bound Frenkel pairs, and before the formation of a dislocation network [6.45]. The corresponding He concentration range is from 10^{-4} to 10^{-2}.

6.5 Swelling of Various Metal Tritides

In a rather old study, the swelling of pressed pellets of yttrium ditritide samples [6.47] was measured using a vernier micrometer. A steady expansion rate was observed over 200 days with a somewhat higher expansion during the first 10 days. The relative length change per generated helium

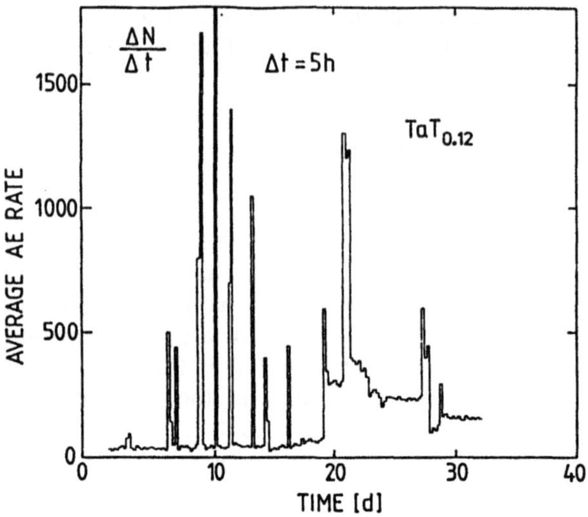

Fig.6.7. Average acoustic emission rate of a $TaT_{0.12}$ sample as a function of aging time [6.45]

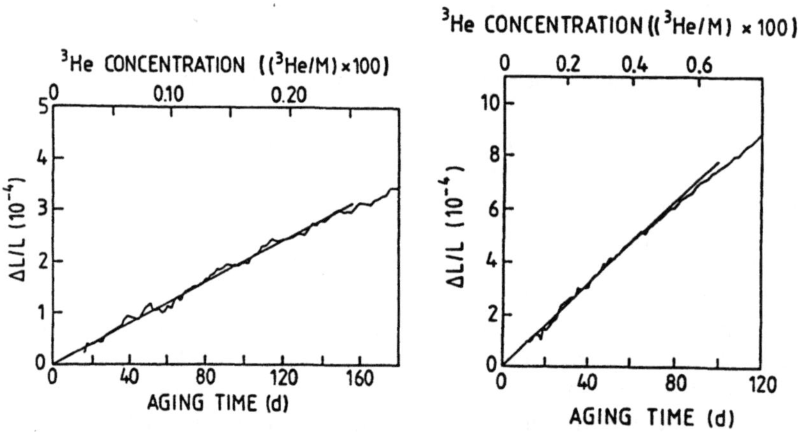

Fig.6.8. Swelling curve of a $TaT_{0.103-c}He_c$ (left-hand side) and a $TaT_{0.42-c}He_c$ (right-hand side) sample as a function of aging time or ^3He concentration at room temperature [6.49]

concentration c can be calculated using [6.47]: $d(\Delta L/L)/dc \simeq 0.1$. *Beavis* and *Miglionico* [6.48] studied thin Er, Ho, Sc, Ti and Y tritide films over a period of 5 to 6 years using interference and scanning electron microscopy. The films of these materials were evaporated onto a polished sapphire substrate covered with a thin layer of chromium. All tritides behaved in a similar way and showed a steady increase in film thickness with ^3He concentration. The rates are similar to the one reported above. After a few years these tritides showed considerable spalling [6.48].

Recently, the swelling rates of different metal tritides were measured at room temperature by *Schober* et al. [6.49] using dilatometric methods. Strain gauges were bonded to the flat surface of metal tritides by means of a special cement. Strains as low as $1\cdot 10^{-6}$ can be measured with this technique [6.50]. The relative length change of two different $TaT_{r_0-c}He_c$ samples with $r_0 = 0.103$ and $r_0 = 0.42$ [6.49] are shown in Fig.6.8 as a function of time and ^3He concention. Although the swelling rates of the two samples are different due to the different He generation rates, the relative length changes ($\Delta L/L$) per atomic He concention are equal: $d(\Delta L/L)/dc = 0.123$ [6.49]. Very interesting results are obtained in the case of $LuT_{0.13-c}He_c$ (Fig.6.9). Here two distinct slopes are observed [6.51]. These two regimes could reflect two stages [6.51] of the self-trapping process [6.9]: interstitial clustering of ^3He, and bubble formation via He-Frenkel-pair generation. Furthermore, the dilation of a $LuT_{0.11-c}He_c$ sample was measured at 4.2 K [6.51].

The relative length change $\Delta L/L$ or volume change $\Delta V/V$ can be described by [6.49]

$$3\frac{\Delta L}{L} = \frac{\Delta V}{V} = c\left(\frac{\Delta v}{\Omega}\right)^{T\rightarrow He} \simeq c\left\{\frac{v_{He}}{\Omega}\frac{\Delta\tilde{v}_I}{\Omega} - \frac{\Delta v_T}{\Omega}\right\}, \tag{6.9}$$

where $\Delta v^{T\rightarrow He}$ is the volume change induced by the decay of a tritium atom, L the length of the sample after tritiding, Ω the host atom volume, v_{He} the volume required by ^3He in the high pressure bubbles, Δv_T the volume change per tritium atom, and $\Delta\tilde{v}_I$ the volume change owing to the

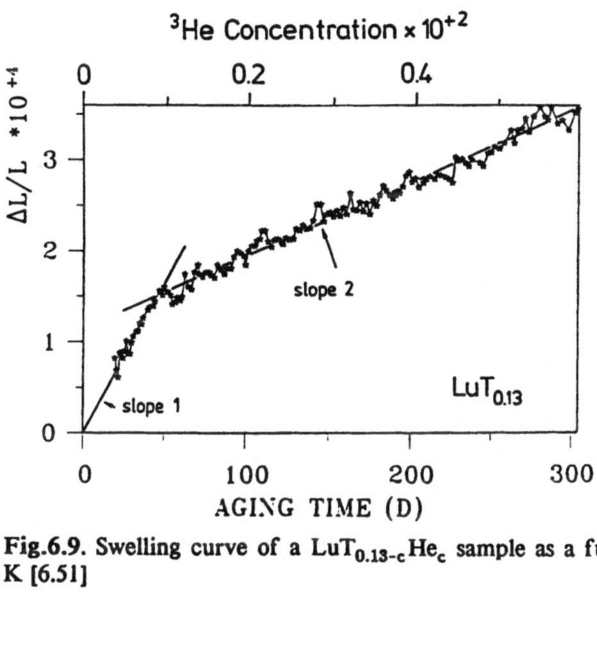

Fig.6.9. Swelling curve of a $LuT_{0.13-c}He_c$ sample as a function of aging time at 293 K [6.51]

Table 6.1. Results of swelling experiments for different metal tritides at room temperature [6.45, 49, 51]

Sample	$(\Delta v/\Omega)^{T\to{}^3He}$	v_{He}/Ω	v_{He} $[10^{-30} m^3]$	p [kbar]
$TaT_{0.103}$	0.37	0.53	9.64	53
$TaT_{0.42}$	0.37	0.52	9.98	47
$NbT_{0.59}$	0.23	0.39	7.7	106
$VT_{0.02}$	0.335	0.52	7.3	12.4
$LuT_{0.13}$				
Slope 1	0.46	0.56	16.7	-
Slope 2	0.123	0.23	6.8	150

transfer of a SIA from a bubble into the lattice, to loops, SIA clusters, dislocations and internal surfaces. c can easily be calculated with (1.2). If the emitted SIAs are incorporated in dislocation loops, $\Delta\tilde{v}_I/\Omega = 1.0$, whereas in the case of isolated SIAs $\Delta\tilde{v}_I/\Omega = 1.1$. The first and second terms of (6.9) describe the volume changes due to the production of a He Frenkel-pair defect and due to the decay of tritium atoms, respectively. An assumption in (6.9) is that all He atoms are contained in bubbles.

In dilatometric experiments, the quantity to be determined is $(\Delta v/\Omega)^{T\to He}$ which allows v_{He} to be calculated. Application of a high-density equation of state for ^3He [6.7] yields very high pressures in the bubble in the range up to 150 kbar. Under these high pressures, it is plausible that ^3He exists as a liquid or a solid inside the bubble at temperatures far above 4.2 K (Sect.6.3) [6.41]. Furthermore, these high pressures are in good agreement with the values required for dislocation loop punching, see (6.8), and with electron energy loss spectroscopy data for ^4He in Al [6.52, 53]. A list of the metal tritides studied to date by *Schober* et al. [6.45,, 51, 54] is given in Table 6.1 with the values for $(\Delta v/\Omega)^{T\to{}^3He}$, v_{He}/Ω, v_{He} and the calculated pressures.

6.6 Lattice Parameter and Width of the Rocking Curves of $TaT_{r_0-c}He_c$ Samples in the α Phase

In the preceding section the volume change upon aging of different metal tritides was presented. A further quantity of interest is the lattice parameter. Both the relative length and lattice parameter changes provide information about the concentration of defects in metals. To date, very little is known about the time dependence of the lattice parameter in metal tritides. In yttrium ditritide a strong increase of the lattice parameter was reported [6.47]. The following lattice parameters were listed [6.47]: 0.5187(2) nm for a new $YT_{2.0}$ sample and 0.5198(2) and 0.5206(2) nm for yttrium ditritide with an aging time of 60 and 150 days, respectively. In contrast, in a later publication by *Beavis* and *Miglionico* [6.48] only very small or no changes at all have been reported for erbium and scandium tritides.

Recently, lattice parameter measurements on $TaT_{r_0-c}He_c$ samples as a function of aging time or increasing He concentration were given using X rays by *Lässer* et al. [6.55, 56] and neutrons by *Blaschko* et al. [6.57, 58].

6.6.1 X-Ray Diffraction Measurements of $TaT_{r_0-c}He_c$

In the X-ray experiment [6.55, 56], a single crystal $TaT_{r_0-c}He_c$ of initial composition $r_0 = 0.164$ was used. Figure 6.10 shows the Bragg peaks of the Ta crystal before and 6, 75 and 492 days, after T charging. The peaks in the centre and lower parts of Fig.6.10 are shifted to smaller Bragg angles due to the absorbed T atoms. The shape of the Bragg peak (circles in Fig.6.10) was measured for the first time six days after tritium charging and was very similar to that of the pure Ta crystal. In contrast, the width of the rocking curve measured 75 days after tritium charging is seen to be larger (squares in the middle part of Fig.6.10). A further dramatic change of shape is seen in the bottom part. Here, only a small portion of the measured Bragg peak is shown using the same scale of the x-axis as in the other plots. All three Bragg peaks of the tritium charged Ta sample appear at about the same Bragg angle. This means that the changes of the averaged lattice parameter upon aging are very small. The rocking curves become asymmetric with increasing age. The distribution of lattice parameters, however, increases. The corresponding values are shown as $\Delta G/G_0$ ($\Delta G/G_0 = -\Delta a/a_0$) versus aging time (lower x-axis) and ^3He concentration (upper x-axis) in the upper part of Fig.6.11. In the lower part the Full Width at Half Maximum (FWHM) of the rocking curve projected on the radius vector of the Ewald sphere $\Delta G_{1/2}/G$ (left-hand y-axis) is plotted as a function of aging time and He concentration. The values on the right were calculated with the relation $\delta G_{1/2}/G = \delta\theta_{1/2}\cos\theta$, where θ is the Bragg angle. The values $(\delta G_{1/2}/G)_{He}$ were determined with a simple devolution procedure [6.56]. The following relations [6.56] were obtained for the relative lattice parameter change $\Delta a/a_0$ and for the relative increase of the full width at half maximum (FWHM) $(\delta G_{1/2}/G)_{He}$ with atomic helium concentration $c (c \leq 0.0015)$:

$$\frac{d\Delta a/a_0}{dc} = (0.6 \pm 1.7)\cdot 10^{-2} \qquad \text{and} \tag{6.10}$$

$$\frac{d(\delta G_{1/2}/G)_{He}}{dc} = 0.21 \pm 0.06 \ , \tag{6.11}$$

where Δa is the difference between the lattice parameter of the $TaT_{0.164-c}He_c$ sample and the lattice parameter a_0 of the unloaded Ta

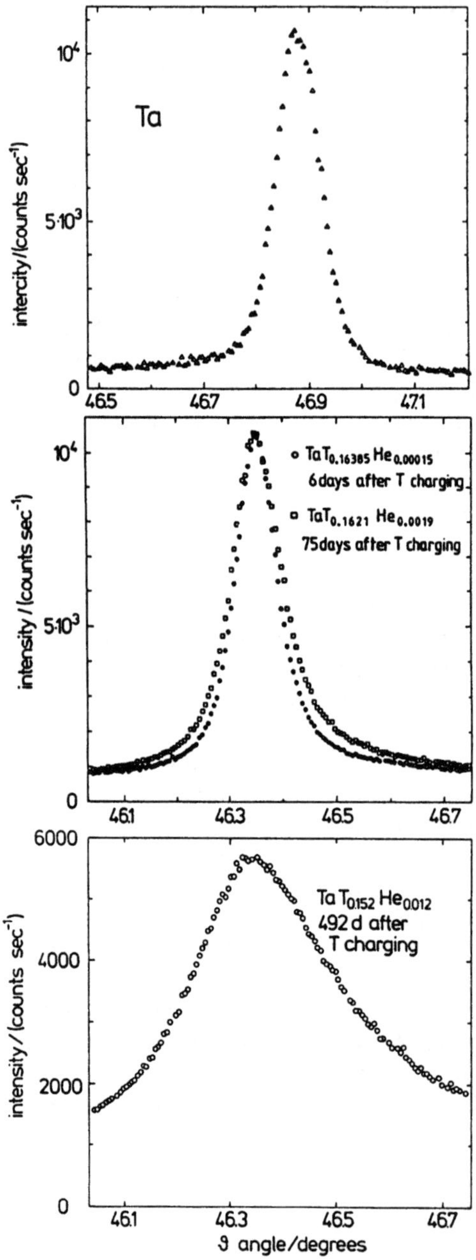

Fig.6.10. Scattering intensity of a Ta single crystal before T loading (upper part), of a $TaT_{r_0-c}He_c$ sample 6 (o) and 75 (□) days (middle) [6.56] and 492 (o) days (lower part) after T charging with the initial T concentration $r_0 = 0.164$ as a function of the scattering angle θ

sample. In the following, the information that can be deduced from the simultaneous length (or volume) and lattice parameter changes is discussed. In previous sections it was pointed out that during the precipitation of ^{3}He in metals, SIAs and dislocation loops are emitted by He clusters and He

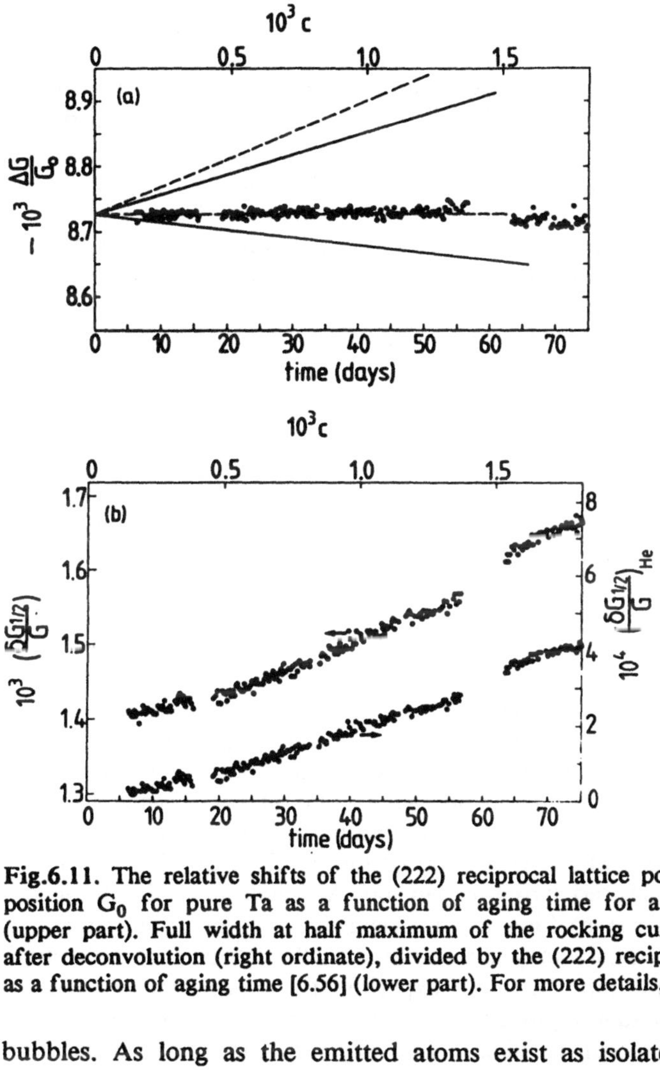

Fig.6.11. The relative shifts of the (222) reciprocal lattice point with respect to its position G_0 for pure Ta as a function of aging time for a $TaT_{0.164-c}He_c$ sample (upper part). Full width at half maximum of the rocking curve (left ordinate) and after deconvolution (right ordinate), divided by the (222) reciprocal lattice parameter as a function of aging time [6.56] (lower part). For more details, see text

bubbles. As long as the emitted atoms exist as isolated SIAs or as very small clusters in the interstitial volume, with a diameter much smaller than the distance to the next nearest cluster, it is expected that 1) the relative length and lattice parameter changes are equal. 2) Diffuse scattering should occur in the vicinity of the Bragg peaks. 3) The width of the Bragg peaks should not change significantly. In contrast, if the emitted SIAs are incorporated into an evolving dislocation network, the relative lattice parameter change is much smaller than the relative length change. The width of the Bragg peaks should increase significantly. The upper two straight lines in Fig.6.11 describe the lattice parameter change assuming only isolated SIAs. They were calculated using the relative length changes presented in the previous section. Here, for the case of isolated SIAs, the relative length and lattice parameter changes are equal. The solid and dashed lines describe the cases where the tritium concentration is constant or decreases due to the radioactive decay, respectively. Constant tritium concentrations can be achieved experimentally by achieving an equilibrium with a higher concentration phase or with the gas phase. In such a case, the loss of tritium due to β^- decay is compensated by the other solid or gaseous phase. The other two straight lines are the result of simple models which neglect any influence of He and consider only the expansion due to tritium. The experimental $\Delta G/G_0$ values lie between the values of the two extreme models. Since the $TaT_{0.164}$ sample contains only one phase, the solid lines should be compared with the experimental data.

According to the famous Simmons-Balluffi equation [6.59]

$$c_V - c_I = 3\left(\frac{\Delta L}{L_0} - \frac{\Delta a}{a_0}\right) \tag{6.12}$$

the concentration $c_V - c_I$ can be calculated with (6.9, 10). c_V is the concentration of vacancies in the bubbles, and c_I the concentration of single SIAs and isolated clusters of SIAs. Using the results of the swelling experiments where $v_{He}/\Omega = 0.52$ or $c_V = 0.52c_{He}$, the following equation is obtained

$$c_V - c_I = 0.36c_{He} \simeq 0.70c_V \tag{6.13}$$

which means that 70% of the emitted SIAs are incorporated into an evolving dislocation network. The rest exist as isolated SIAs or in very small, isolated clusters of SIAs. These 30% of SIAs are the main cause of the increase of the lattice parameter upon aging. This effect somewhat overcompensates for the shrinkage of the lattice due to T loss. The lattice parameter would decrease as shown by the bottom straight line in Fig.6.11, if there were no He-induced lattice expansion. A simple analysis of the increasing FWHM of the rocking curves yielded $c_V - c_I \simeq 0.75c_V$ [6.56], a result which was in satisfactory agreement with the lattice parameter analysis. Therefore, combining different experiments, a consistent explanation of both the peak broadening and the relative change of the lattice parameter and of the volume can be given. It is very surprising that a dislocation network is built up prior to the beginning of the experiment which is within 6 days of charging.

Figure 6.12 shows the relative FWHM of the (222) Bragg peak measured over a period of 500 days. For the measured He concentration of up to 1.2% no saturation of the width increase is seen. This indicates that the dislocation density is still increasing.

6.6.2 Neutron-Scattering Measurements of $TaT_{r_0-c}He_c$

In the neutron-scattering experiment [6.57] two polycrystalline tantalum platelets ($50 \cdot 50 \cdot 1\,mm^3$) were charged with tritium to initial concentrations: $r_0 = 0.04$ and $r_0 = 0.115$. A lattice parameter increase and an increase of

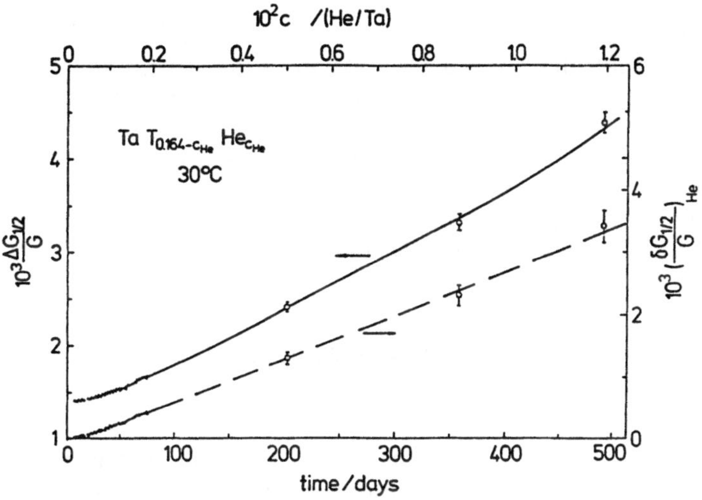

Fig.6.12. Full width at half maximum of the experimental rocking curve divided by the (222) reciprocal lattice parameter as a function of aging time for $TaT_{0.164-c}He_c$

the FWHM of the Debye-Scherrer lines were observed. *Blaschko* et al. [6.57] reported the following lattice parameter increase

$$\frac{d(\Delta a/a_0)}{dc} = 0.102 \pm 0.040 \ . \tag{6.14}$$

Within experimental accuracy this value agrees with the value determined by the swelling experiment of *Schober* et al. [6.49] ($d(\Delta L/L_0)/dc_{He} = 0.127$) indicating that only a very small part of the SIAs would be transferred to dislocations. Starting from this assumption, it is difficult to explain the observed broadening of the Debye-Scherrer lines. Possible reasons for the different quantitative results [6.56, 57] could be: With X rays, only a few micronmeters at the surface are observed, whereas neutrons can easily penetrate thick samples. In the X-ray experiment, a single crystal with an initial concentration of $r_0 = 16.4\%$ was used. In the neutron work polycrystalline samples with lower initial concentrations were employed. Furthermore, the polycrystalline samples disintegrated at low He concentrations ($c_{He} \simeq 0.5\%$ or 0.9%) probably due to He embrittlement of grain boundaries. The single crystal employed for the X-ray study [6.56], is still intact, although the He concentration exceeded 1.4%.

In addition, *Blaschko* et al. [6.57] found an anisotropic broadening of different Debye-Scherrer lines for the polycrystalline $TaT_{0.06-c}He_c$ sample and studied long wavelength phonons in a $TaT_{0.05-c}He_c$ crystal. In comparison to pure Ta a reduction of the phonon intensities for $TaT_{0.05-c}He_c$ was observed.

Further work is necessary to clarify the evolution of the lattice parameter in metal tritides. Simultaneous measurement of the lattice-parameter and length changes would be very useful. In addition, experiments at very low temperatures should allow the determination of the lattice expansion due to interstitially dissolved He because the He atoms are immobile at very low temperatures (Sect.6.7.4).

On the other hand, the above mentioned observations of the volume and lattice-parameter changes in metal tritides mean that their interpretations are much more difficult than for MD_r or MH_r. Only for young MT_r samples can similar conclusions as for MH_r or MD_r be drawn.

6.7 Preliminary Results for Helium-3-Induced Effects in Selected Metals

In the following sections short descriptions of a few results obtained very recently in this laboratory and in collaboration with other groups will be given.

6.7.1 Mechanical Properties of Metal Tritides

It is well known that the mechanical properties of metals are changed by the presence of hydrogen and helium atoms [6.16]. Hydrogen embrittlement and intergranular fracture due to He precipitation at grain boundaries play an important role in determining the stability of structural materials.

Caskey et al. studied the mechanical behaviour of $MT_{r_0-c}He_c$ alloys [6.40]. Effects due both to T and He were considered as well as possible He-T interactions.

Schober et al. [6.60] measured the Vickers hardness of $NbT_{r_0-c}He_c$ and $TaT_{r_0-c}He_c$ samples applying a load of 200 g. As an example, the Vickers hardness of a $TaT_{0.097-c}He_c$ sample is shown in Fig.6.13 as a function of aging time. The strong increase is produced by the generation of ^3He filled bubbles and an evolving dislocation network.

6.7.2 He-Induced Change of the Tritium Diffusion in Dilute Tantalum Tritides

The diffusivity of metal tritides was studied at room temperature by *Schober* et al. [6.60]. They used the Gorsky effect as measured with strain gauges [6.50]. Their diffusion constant for a young $TaT_{0.049}$ sample was in good agreement with the data of *Qi* et al. [6.61] (Sect.5.10.1). A slight decrease of the diffusivity as a function of aging time can easily be seen in Fig.6.14 despite the large scatter of the data. A simple model neglecting He-induced effects would yield the opposite result: an increasing diffusivity due to the smaller blocking probability of nearest neighbours. The observed decrease of the diffusion constant can be attributed to the He-induced damage of the host lattice which causes trapping of T atoms.

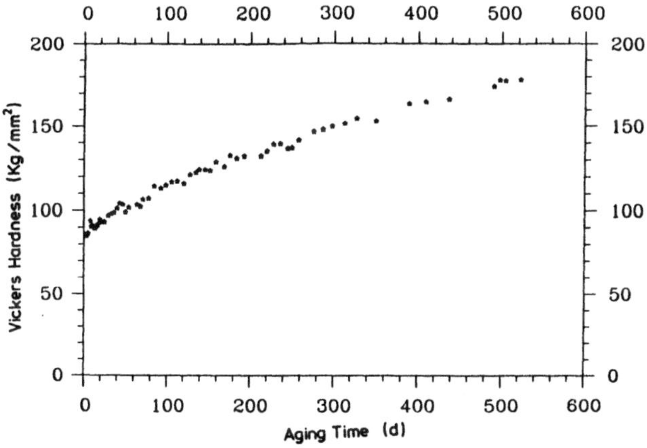

Fig.6.13. Vickers hardness of a $TaT_{0.097-c}He_c$ sample [6.60] versus aging time

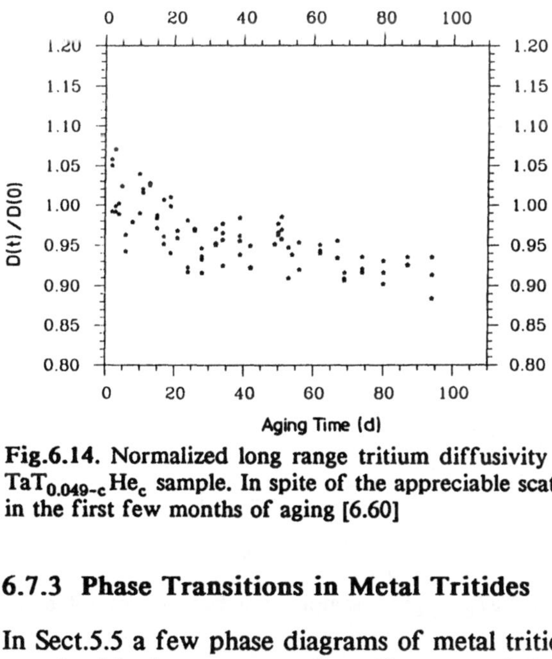

Fig.6.14. Normalized long range tritium diffusivity D(t)/D(o) versus aging time in a $TaT_{0.049-c}He_c$ sample. In spite of the appreciable scatter a continuous drop is observed in the first few months of aging [6.60]

6.7.3 Phase Transitions in Metal Tritides

In Sect.5.5 a few phase diagrams of metal tritides were presented and compared with the corresponding diagrams of the metal protides and deuterides. In the case of M-T phase diagrams, the data were obtained using relatively young samples.

Figure 6.15 shows the differential thermal analysis (DTA) signals of a $TaT_{0.691-c}He_c$ sample in the temperature range 50° to 75°C as a function of aging time [6.62]. The phase boundaries are shifted to higher temperatures compared with H and D. No phase diagram of Ta-T has been published as yet. Assuming that the Ta-T and Ta-D phase diagrams are topo-

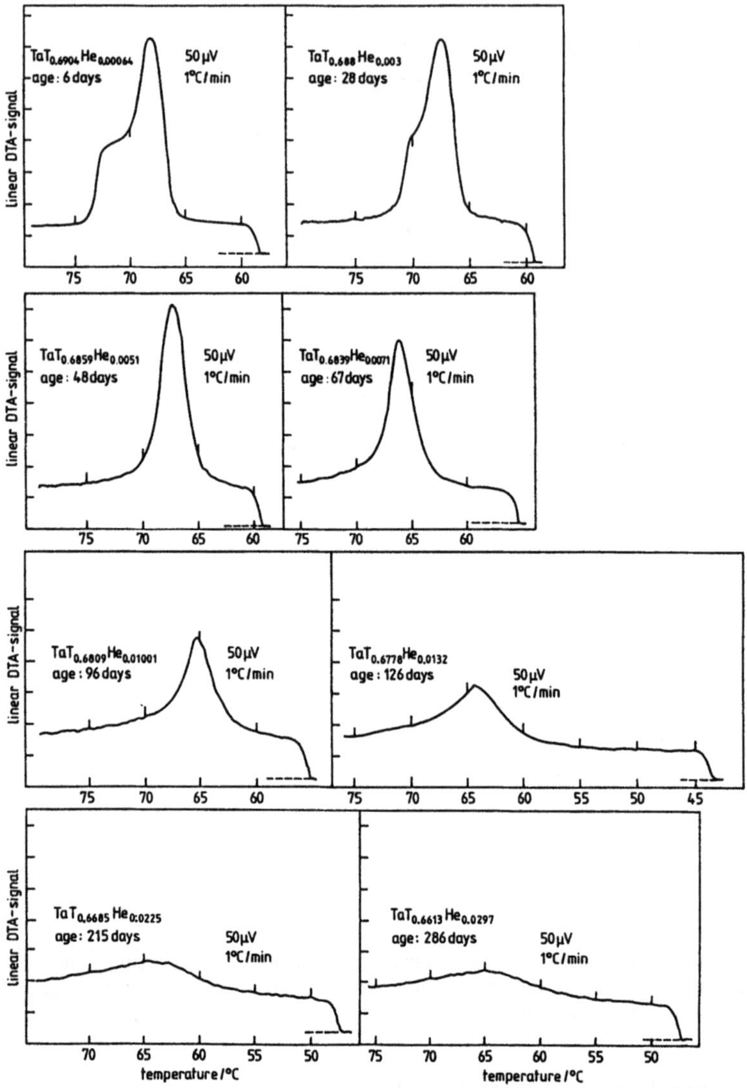

Fig.6.15. Linear DTA signal of a $TaT_{0.691-c}He_c$ sample as a function of different aging times [6.62]

logically equivalent for concentrations between 0.65 and 0.70, the dramatic decrease of the reaction enthalpies of the phase transitions as a function of age cannot be explained by lower tritium concentrations alone. A possible explanation may be that, due to the growing amount of He, the structure of the ordered metal tritides becomes partially disordered, resulting in a smaller enthalpy contribution (Fig.6.15). The temperature shifts of the peaks correlate well with the expected and extrapolated shifts of the Ta-D system indicating that the sample has not lost tritium.

6.7.4 Resistivity of Lutetium Tritides

Resistivity measurements of $LuT_{r_0-c}He_c$ ribbon [6.63] were performed to obtain information on diffusion and resistivity of ^3He in Lu. Lu was chosen because of its very high solubility limit of about 20% in the α phase even at very low temperatures, which is due to a temperature-independent solvus line. This allows LuT_r samples with $r \le 0.20$ to be cooled down to liquid He temperatures without precipitation of high concention phases. In addition, at 4.2 K the ^3He atoms appear to be frozen-in preventing ^3He cluster and bubble formation. The following observations were made [6.63]: 1) A similar resistivity anomaly as found in LuH_r and LuD_r [6.64] is also observed in LuT_r. The temperature shift of the anomaly followed the trend expected from the known behaviour of H and D in Lu. This resistance anomaly is caused by pairing or clustering of hydrogen atoms during cooling. 2) An increase of the resistivity of $LuT_{r_0-c}He_c$ samples with the initial tritium concention r_0 in the α phase was observed as a function of aging time at 4.2 K. The resistivity increment per He atom was found to be about twice that of T which was the first determination of the resistivity of presumably interstitially dissolved He in a metal. 3) Annealing experiments after aging at 4.2 K show that the ^3He atoms migrate in a temperature range above about 26 K. The recovery of the resistivity depended strongly on the cooling rate.

6.7.5 Analysis of Metal Tritides Using Positron-Annihilation Techniques

Positrons annihilate in condensed matter preferentially at defects such as vacancies and voids where they are easily trapped. Positron annihilation is an important tool to study micrOscopic defects and defect clusters, where many other techniques fail.

Positron lifetimes and Doppler broadening measurements of tritiated metals have been performed [6.65]. Figure 6.16 shows results for $NbT_{0.028-c}He_c$. For all He concentions two distinct lifetimes τ_1 and τ_2 with intensities I_1 and I_2, respectively, have been observed. The longer lifetime τ_2 is due to annihilations in the defects. In the early stage of He segregation τ_2 decreases strongly. After ~70 days, τ_2 reached a constant value of about 195 ps, which is too low to be explained by He-filled bubbles. A possible explanation may be that some positrons are trapped at dislocations or dislocation loops generated by the bubbles. The characteristic lifetime for annihilation at dislocations is about 170 ps [6.65]. On the other hand, no lifetimes were found which can be attributed to the bubbles observed by TEM in other materials.

In addition to τ_2, Fig.6.16 shows the lifetime τ_b of positrons annihilating in the bulk of the host metal and the total trapping rate K of the defects.

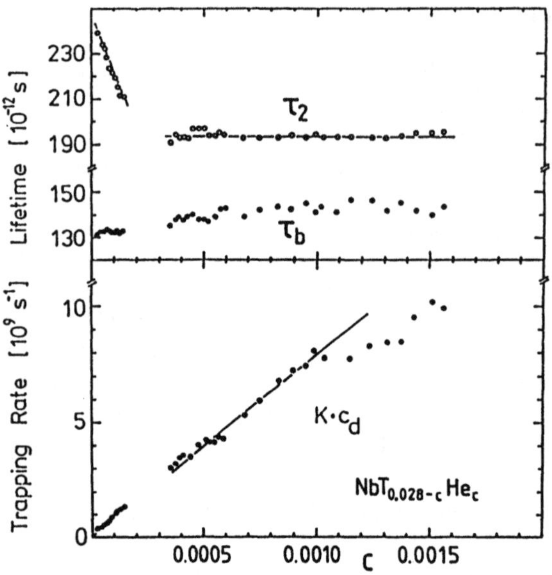

Fig.6.16. Positron lifetime and trapping rate of a $NbT_{0.028-c}He_c$ sample as function of increasing He concentration [6.65]

Both quantities are derived from the two-state trapping model [6.66] by means of the following relations

$$K = I_2 \frac{\tau_2 - \tau_1}{\tau_1 \tau_2} , \tag{6.15}$$

$$\tau_b = \frac{\tau_1}{1 - K\tau_1} . \tag{6.16}$$

The trapping rate K is proportional to the He concentrations for $c \leq 0.001$. The increase of τ_b with increasing He concentration indicates an expansion of the lattice. To clarify these observations, further studies - also of He-implanted and deformed materials - are in progress.

6.7.6 Room-Temperature Creep of Metal Tritides

In Sect.6.5, linear swelling of bulk metal tritides is discussed. That work has recently been extended to cases where the swelling metal tritide was under an uniaxial stress almost amounting to the yield stress of the metal. In dilute tantalum tritides, a new strain rate was observed which was roughly 2 to 3 times higher than the rate in the unstressed tritide [6.67]. This phenomenon is referred to as room temperature creep of the tritide; it is analogous to the irradiation creep also occurring at low temperatures (note that regular creep occurs at temperatures above $\sim T_m/3$, where T_m is the melting point in Kelvin).

The present low-temperature creep may be modelled in a crude, first approximation as follows [6.67]: it is supposed that all SIA loops are punched out along favourable ⟨hkl⟩ directions. These are directions that are roughly parallel with the stress axis. In addition, it is assumed that most single SIA atoms cluster in such a way as to effectively elongate the sample along the stress axis. Under these conditions

$$\dot{\epsilon}_{creep} \simeq 3\dot{\epsilon}_{swelling} , \tag{6.17}$$

where $\dot{\epsilon}_{swelling}$ is the swelling rate in the unstressed state.

7. Conclusion and Outlook

An attempt has been made to address many different tritium-related topics, for instance, production of tritium in the atmosphere, in reactors, in nuclear explosions and in future fusion reactors, risks of tritium, release of tritium to the environment, ways of doping metals with tritium and the detection of tritium in metals. In addition, various aspects of T and ^{3}He in $MT_{r_0-c}He_c$ have been reviewed.

In recent years, several physical quantities have been determined for the first time for tritium in metals. However, the information about T in metals is still scarce in comparison to that for H and D. Simple extrapolation of the known behaviour of H and D may result in large errors of the corresponding values for T if the H and D data were determined by different groups. Different systematic experimental errors may often hide the correct isotope effect. Therefore, isotopic dependence was only discussed for those properties that were measured under the same experimental conditions.

In the case of H, D and T in metals, the isotope dependence of a few physical quantities can be explained qualitatively by the mass dependence of the optical phonons. But even in the case of the very dilute Pd-X (X = H, D, T) system there is no physical model to quantitatively describe the solubility using the partition function and the optical mode energies determined by inelastic neutron scattering experiments. Furthermore, to date, the optical phonons of tritium in metals have only been determined for the β phase of niobium and palladium[1]. Due to unknown anharmonic effects and changes of the electronic configuration, a quantitative description of the solubility, or of phase diagrams, is obtained only in very few cases. The behaviour of tritium in metals can often be predicted in a qualitative way from the known properties of the stable hydrogen isotopes. Exceptions are, for example, the concentration dependence of special extremal cross sections of the Fermi surface in the Pd-X (X = H, D, T) system and the diffusion of the hydrogen atoms at low temperatures.

The absolute shifts of the following physical properties in metals due to tritium were reviewed and compared with those of H and D: solubility at low and/or high hydrogen concentrations in Pd, $Pd_{1-z}Ag_z$, Li, Y and V, phase diagrams of the Pd-X, V-X and Nb-X (X = H, D, T) systems, optical vibrations of the hydrogen isotopes in Nb and Pd, superconductivity, diffusion, etc. Isotope studies may be the only way to confirm a theory, if

1 During publication, a further paper appeared discussing the optical phonons of tritium in the β-phase of vanadium (J.J. Rush, N.F. Berk, A. Magerl, J.M. Rowe, J.L. Provo: Phys. Rev. B 37, 7901 (1988))

different models can describe the same temperature dependence, as mentioned for the diffusion of H, D, T in metals.

In addition, the properties of the tritium decay product ^{3}He were also reviewed, with the emphasis on recent results. The following effects of ^{3}He in metals were discussed: He retention in, and release from, metals, observation of ^{3}He bubbles, punching out of dislocation loops, emission of acoustic sound waves, swelling, change of lattice parameter, migration, mechanical properties and resistivity. The generation of bubbles can be explained consistently by the self-trapping process where He atoms may generate bubbles in the absence of lattice defects by the emission of self-interstitial atoms. So far, mainly the separate influence of either T or He in metals has been discussed in most cases, neglecting possible interactions between T and He. It is often assumed that the observed He-induced changes in metal tritides can be generalized for the pure metal; the dissolved tritium atoms are considered merely as an impurity necessary for the generation of He, but as having no marked influence on the metal properties. In many cases this assumption is an oversimplification, especially for cases where the host lattice structure changes upon tritium absorption. On the other hand, the increasing He concentration may also influence the physical properties of tritium in the metal. This is clearly seen in the swelling, lattice-parameter and diffusivity experiments. It means that in order to study purely tritium-induced effects, metal tritides of very young age have to be used.

Due to growing safety precautions in working with radioactive materials, and the use of kilogram quantities of tritium in future fusion power reactors, metals will be employed more frequently for handling large amounts of tritium. According to the specifications of different applications, metal tritides can be found which are very suitable materials for the reversible and final storage, pumping, compression, purification, isotope enrichment, transport and waste disposal of tritium.

The further determination of the unknown properties of tritium and its decay product ^{3}He in metals is of great importance from a basic research standpoint as well as for the technological application of these alloys.

References

Chapter 1

1.1 L.W. Alvarez, R. Cornog: Phys. Rev. **56**, 613 (1939)
1.2 W.F. Libby: Phys. Rev. **69**, 671 (1946)
1.3 M.L. Oliphant, P. Harteck, Lord Rutherford: Nature **133**, 413 (1934)
1.4 W.F. Libby: In *Tritium*, ed. by A.A. Moghissi, M.W. Carter Conf.710809, Las Vegas, 1973
1.5 W.F. Libby: In *Tritium in the Physical and Biological Sciences I* (International Atomic Energy Agency, Vienna 1962) p.3
1.6 E.A. Evans (ed.): *Tritium and its Compounds* (Butterworth, London 1974)
1.7 P. Hautojärvi (ed.): *Positrons in Solids*, Topics Curr. Phys., Vol.12 (Springer, Berlin, Heidelberg 1979)
1.8 E. Karlsson: Phys. Rep. **82**, 271 (1982)
1.9 D. Richter: In *Neutron Scattering and Muon Spin Rotation*, Springer Tracts Mod. Phys. **101**, 85 (1983)
1.10 A. Schenck (ed.): *Muon Spin Rotation Spectroscopy* (Hilger, Bristol 1985)
1.11 A. Seeger: In *Hydrogen in Metals I*, ed. by G. Alefeld, J. Völkl, Topics Appl. Phys. Vol.28 (Springer, Berlin, Heidelberg 1978) p.349
T. Springer: In *Hydrogen in Metals I* ed. by G. Alefeld, J. Völkl, Topics Appl. Phys. Vol.28 (Springer, Berlin, Heidelberg 1978) p.75
1.12 J.P. Bloxsidge, J.A. Elvidge: Progress in NMR Spectroscopy **16**, 99 (1983)
1.13 E.A. Evans, D.C. Warrell, J.A. Elvidge, J.R. Jones (eds.): *Handbook of Tritium NMR Spectroscopy and Applications* (Wiley, New York 1985)
1.14 F.A. Lewis (ed.): *The Palladium Hydrogen System* (Academic, London 1967)
W.M. Müller, J.P. Blackledge, G.G. Libowitz (eds.): *Metal Hydrides* (Academic, London 1968)
E.L. Muetterties (ed.): *Transition Metal Hydrides* (Marcel Dekker, New York 1971)
G. Alefeld, J. Völkl (eds.): *Hydrogen in Metals I* and *II*, Topics Appl. Phys. Vols.28,29 (Springer Berlin, Heidelberg 1978)
E. Fromm, E. Gebhardt (eds.): *Gase und Kohlenstoff in Metallen* (Springer, Berlin, Heidelberg, 1976)
H. Buchner (ed.): *Energiespeicherung in Metallhydriden* (Springer, Wien 1982)
J.W. Ward: In *Handbook on the Physics and Chemistry of the Actinides*, ed. by A.J. Freeman, C. Keller (Elsevier, Amsterdam 1985) Chap.1
L. Schlapbach (ed.): *Hydrogen in Intermetallic Compounds*, Vol.I, Topics Appl. Phys. Vol.63 (Springer, Berlin, Heidelberg 1988)
1.15 P. Jena, C.B. Satterthwaite (eds.): *Electronic Structure and Properties of Hydrogen in Metals* (Plenum, New York 1983) p.1
W.E. Wallace, T. Schober, S. Suda (eds.): *Metal Hydrides* 1982, Proc. Int'l Symp. on the Properties and Applications of Metal Hydrides, Toba, Japan 1982 (Elsevier Sequoia, Lausanne 1983), Vols.I,II; see also J. Less-Common Metals **88**, and **89** (1982)

F.A. Lewis, E. Wicke (eds.): Hydrogen in Metals. Int'l Conference, Wroclaw, Poland (1983); see also J. Less-Common Metals **101** (1984)
M. Ron, Y. Josephy, I. Jacob, D. Shaltiel (eds.): *Metal Hydrides*, 1984, Proc. Int'l Symp. on the Properties and Applications of Metal Hydrides IV, Eilat Israel, 1984 (Elsevier Sequoia, Lausanne 1985) Vols.I,II; see also J. Less-Common Metals **103** and **104** (1984)
F.A. Lewis, E. Wicke (eds.): Hydrogen in Metals. Int'l Conf., Belfast, Northern Ireland (1985), see also Z. Phys. Chem., NF, **143** (1985) and **147** (1986)
G. Bambakidis, R.C. Bowman, Jr (eds): *Hydrogen in Disordered and Amorphous Solids*, NATO ASI Series, Series B: Physics. Vol.136 (Plenum, New York 1985)
A. Percheron-Guegan, M. Gupta (eds.): Metal Hydrides 1986, Proc. Int'l Symp. on the Properties and Applications of Metal Hydrides V, Maubuisson, France (1986); see also J. Less-Common Metals **129** and **130** and **131** (1987)
E. Fromm, R. Kirchheim (eds.): Int'l Symp. on Metal Hydrogen Systems, Fundamentals and Applications, Stuttgart, 4-9 Sept. 1988, to be published
1.16 J.F. Watson, F.W. Wiffen (eds.): Proc. Int'l Conf. Radiation Effects and Tritium Technology for Fusion Reactors, Gatlinburg, TN, October 1-3 (1975), CONF-750989 (National Technical Information Service, Springfield, 1976) Vols.1-4
L.J. Wittenberg (ed.): Proc. Tritium Technology in Fission, Fusion and Isotopic Applications, American Nuclear Society National Topical Meeting, Dayton, OH (1980), U.S. Department of Energy CONF-800 427 (Southwestern Ohio Section, American Nuclear Society, 1981)
Proc. 14th Symp. on Fusion Technology 1986 (Pergamon, London)
Proc. Tritium Technology in Fission, Fusion and Isotopic Applications, Toronto, Canada (1988), to be published
Proc. 15th Symp. on Fusion Technology, Utrecht, The Netherlands (1988), to be published
J. Raeder, K. Borrass, R. Bünde, W. Dänner, R. Klingelhöfer, L. Lengyel, F. Leuterer, M. Söll (eds.): *Kontrollierte Kernfusion* (Teubner, Stuttgart 1981)
1.17 C.R. Ridu, K.C. Jordan: Mound Laboratory Report MLM-2458, Miamisburg, Ohio 54342 (1977)
1.18 P.C. Souers ed.): *Hydrogen Properties for Fusion Energy* (Univ. California Press, Berkeley 1986)
1.19 R. Lässer, G.L. Powell: Phys. Rev. B **34**, 578 (1986)

Chapter 2

2.1 UNSCEAR. Ionizing Radiation: Sources and Biological Effects (United Nations, New York 1982)
2.2 F.J.L. Bindon: Nucl. Eng. 27, 52 (1986)
2.3 F. Luykx, G. Fraser: Radiation Protection Dosimetry **16**, 31 (1986)
H.T. Peterson, Jr., D.A. Baker: Fusion Technology **8**, 2544 (1985)
2.4 F. Luykx, G. Fraser: European Seminar on the risks from tritium exposure, Mol 1982, report EUR 9065 (1982)
2.5 J.A. Miskel: In *Tritium*, ed. by A.A. Moghissi, M.W. Carter, Conf. 710809, Las Vegas, Nevada (1973)
2.6 M.S. Terpilak: Radiological Health Data and Reports 12, 180 (1971)
2.7 M. Eisenbud: Tritium in the Environment, NCRP report No.62 (1978)
2.8 E. Teuchert, private communication.
2.9 Int'l Commission on Radiological Protection "Limits for intakes of radionuclides by workers" ICRP-Publ. 30, Part 1, (Pergamon, London 1978)

2.10 Verordnung über den Schutz vor Schäden durch ionisierende Strahlen (Strahlenschutzverordnung StrlSchV), (BGBl I, 2905; BGBl I, 1977, pp.184,269) 13.10.1976
2.11 D.C. Lloyd, A.A. Edwards, J.S. Prosser, A. Auf der Maur, A. Etzweiler, U. Weickhardt, U. Gössi, L. Geiger, U. Noelpp, H. Rösler: Radiation Protection Dosimetry **15**, 191 (1986)
2.12 R.M. Brown: Fusion Technology **8**, 2539 (1985)
2.13 G.L. Kulcinski, J.M. Dupony, E. Ishino: J. Nucl. Mater. **141–143**, 3 (1986)
2.14 H. Djerassi, J.L. Rouyer: Radiation Protection Dosimetry **16**, 37 (1986)
2.15 P.J. Dinner, W. Gulden: Radiation Protection Dosimetry **16**, 39 (1986)
P.J. Dinner, S.K. Sood, M. Chazalon, O.K. Kreton: Fusion Technology **8**, 2228 (1985)
P. Rocco, R. Kirchmann: Radiation Protection Dosimetry **16**, 49 (1986)
2.16 V.A. Maroni, E.H. van Deventer: J. Nucl. Mater. **85** and **86**, 257 (1979)
2.17 G. Casini, P. Rocco, Ponti: Technical Note No. I.04.B1, 85.156 ISPRA, Italy (1985)
O. Edlund: Radiation Protection Dosimetry **16**, 27 (1986)
P. Rocco, R. Kirchmann: Radiation Protection Dosimetry **16**, 49 (1986)
2.18 R. Lässer, K.-H. Klatt, D. Triefenbach, J.L. Hemmerich: J. Nucl. Mater. **148**, 145 (1987)
2.19 H. Weichselgartner: Vakuum-Technik **34**, 236 (1986)
2.20 J.L. Hemmerich: J. Vac. Sci. Technol. **A6**, 144 (1988)
2.21 J.L. Anderson, D.O. Coffin, C.R. Walthers: J. Vac. Sci. Technol. **A1**, 849 (1983)
2.22 H. Weichselgartner, H. Frischmuth, A. Stimmelmayr, A. El-Sharnouby: Proc. 13th SOFT, Varese (1984) p.441
H. Frischmuth, J. Perchermeier, A. Stimmelmayr, H. Weichselgartner: In *Proc. 12th Symp. on Fusion Technology (SOFT 1982)* Vol.1 (Pergamon, London 1983) p.549
2.23 R.C. Jones, W.H. Furry: Rev. Mod. Phys. **18**, 151 (1946)
W.M. Rutherford, C.N. Lindsag: Fusion Technology **8**, 2278 (1985)
2.24 P.C. Stangeby, J.E. Allen: J. Appl. Phys. **54**, 14 (1983)
2.25 T. Tanifuji, K. Noda, S. Nasu, K. Uchida: J. Nucl. Mater. **95**, 108 (1980)
2.26 A. Ohara, Y. Sakuma, M. Okamoto: J. Nucl. Mater. **102**, 356 (1981)
2.27 K. Kurasawa, H. Watanabe, G.W. Hollenberg, Y. Iskii, A. Nishimura, H. Yoshida, Y. Naruse, M. Aizawa, H. Ohno, S. Konishi: J. Nucl. Mater. **141–143**, 265 (1986)
2.28 J.M. Miller, S.R. Bokwa, R.A. Verall: J. Nucl. Mater. **141–143**, 294 (1986)
2.29 D.L. Baldwin, G.W. Hollenberg: J. Nucl. Mater. **141–143**, 305 (1986)
2.30 C.H. Wu: J. Nucl. Mater. **114**, 30 (1983)
2.31 M. Hoch: J. Nucl. Mater. **120**, 102 (1984)
2.32 M. Saeki, M. Nakashima, Y. Aratono, E. Tachikawa: J. Nucl. Mater. **120**, 267 (1984)
2.33 E. Roth, F. Botter, M. Briec, M. Rostaing, H. Werle, R.G. Clemmer: J. Nucl. Mater. **141–143**, 275 (1986)
2.34 H. Kwast, R. Conrad, P. Kennedy, A.J. Flipot, J.D. Elen: J. Nucl. Mater. **141–143**, 300 (1986)
2.35 M. Briec, F. Botter, J.J. Abassin, R. Benoit, P. Chenebault, M. Masson, B. Rasneur, P. Sciers, H. Werle, E. Roth: J. Nucl. Mater. **141–143**, 357 (1986)
2.36 K. Noda, M. Arita, Y. Ishii, H. Saka, T. Imura, K. Kuroda, H. Watanabe: J. Nucl. Mater. **141–143**, 353 (1986)
2.37 M.C. Billone, Y.Y. Liu, R.B. Poeppel, J.L. Routbort, K.C. Goretta, D.S. Kuppermann: J. Nucl. Mater. **141–143**, 282 (1986)

2.38 P.J.D. Kind, P. Kupschus, G.W. O'Hara (eds.): JET Joint Undertaking, Annual Report 1984, EUR 10222 EN, EUR-JET-AR7 (August 1985)
2.39 JET Joint Undertaking Progress Report 1987, EUR 11579 EN, EUR-JET-PR5 (March 1988)
2.40 J.P. Coad, R. Behrisch, H. Bergsaker, J. Ehrenberg, B. Emmoth, J. Partridge, G. Saibene, R. Sartori, J.C.B. Simpson, Wang Wen Min: To be published in J. Nucl. Mater. 1988
2.41 R. Hancox: To be published
2.42 A.H. Dombra, M. Carney: In Proc. 13th Symp. of Fusion Technology (SOFT 1984) p.473
2.43 W.J. Holtslander, R.E. Johnson, F.B. Gravelle, C.M. Shultz: Fusion Technol. **10**, 1340 (1986)
2.44 R.D. Penzhorn, M. Glugla: Fusion Technol. **10**, 1345 (1986)
2.45 R. Lorscheider: In Herkunft, Handhabung und Verbleib von Tritium, RSI1-510 321/196-SR 165 (1980)
2.46 Handling of Tritium-Bearing Wastes, Techn. Rept. Series No.203, IAEA, Vienna (1981), ISBN 92-0-125081-9
2.47 Management of Tritium at Nuclear Facilities, Techn. Rept. Series No.234, IAEA, Vienna (1984), ISBN 92-0-125084-3
2.48 H. Brücher, E. Merz: Kernforschungsanlage Jülich, JÜL-2099 (1986), ISSN 0366-0885
2.49 H. Brücher, K. Hartmann: Kernforschungsanlage Jülich, JÜL-1838 (1983), ISSN 0366-0885
2.50 P.H. Brücher: Radioactive Waste Management and the Nuclear Fuel Cycle **7**, 195 (1986)
2.51 W.J. Holtslander, J.M. Yaraskavitch: Atomic Energy of Canada Ltd. AECL-7151 (1981)
2.52 W.J. Holtslander, J.M. Yaraskavitch: Atomic Energy of Canada Ltd. AECL-7757 (1982)
2.53 N. Ebeling: Kernforschungsanlage Jülich, JÜL-2010 (1985), ISSN 0366-0885
2.54 K. Krejci, A. Zeller, Jr.: In *Behaviour of Tritium in the Environment*, Proc. Symp. San Francisco, 1978 (IAEA, Vienna 1979) p.65
2.55 G. Wehner: In *Behaviour of Tritium in the Environment*, Proc. Symp. San Francisco, 1978 (IAEA, Vienna 1979) p.79
2.56 F. Combs, R.J. Doda: In *Behaviour of Tritium in the Environment*, Proc. Symp. San Francisco, 1978 (IAEA, Vienna 1979) p.93
2.57 M. Damiani, R. Getraud, A. Senn: Sonderheft NUCLEX 1972, p.41
P.H. Pautrot, M. Damiani: in *Separation of Hydrogen Isotopes*, ed. by H.K. Rae, American Chemical Society Symp. Ser. No.68 (1978)
2.58 G. Vasaru (ed): *Separarea Tritiului* (Editura Dacia, Cluj-Napoca, 1987)

Chapter 3

3.1 H.Wenzl, J.-M. Welter: In *Current Topics Mater. Sci.* **1**, 603 (North-Holland, Amsterdam 1978)
3.2 B. Jungblut, G. Sicking: J. Less-Common Metals **101**, 373 (1984)
J. Chene: J. Less-Common Metals **131**, 337 (1987)
3.3 S. Okuda, R. Taniguchi, M. Fujishiro, Y. Satoh, E. Hiraoka: J. Nucl. Mater. **128-129**, 725 (1984)
3.4 G.U. Greger, H. Münzel, W. Kunz, A. Schwierczinski: J. Nucl. Mater. **88**, 15 (1980)

W. Kunz, H. Münzel, U. Helfrich: J. Nucl. Mater. **105**, 178 (1982)
3.5 R. Lässer, K.-H. Klatt, P. Mecking, H. Wenzl: Kernforschungsanlage Jülich, JÜL-1800 (1982), ISSN 0366-0885
3.6 R. Lässer: In *Gase in Metallen*, ed. by D. Hirschfeld (Deutsche Gesellschaft für Metallkunde e.V. 1984) p.65, ISBN 3-88355-086-8
3.7 P. Mecking-Schoetensack: Jül-1779, Kernforschungsanlage Jülich (1982), ISSN 0366-0885
3.8 K.-H. Klatt, S. Pietz, H. Wenzl: Z. Metallkd. **69**, 170 (1978)
3.9 H. Wenzl: Int'l Metals Rev. **27**, 140 (1982)
3.10 R.S. Carlson: In Proc. Int'l Conf. Radiation Effects and Tritium Technology for Fusion Reactors, Vol.IV, ed. by J.S. Watson, F.W. Wiffen (Gatlinburg 1975) p.36
3.11 R. Lässer, D. Triefenbach, K.-H. Klatt: Jül-Spez.-431, Kernforschunganlage Jülich (1988), ISSN 0343-7639
3.12 R. Lässer: Nucl. Instr. Meth. **215**, 467 (1983)
3.13 H.F. Dylla: J. Vac. Sci. Technol. **20**, 119 (1982)
R.E. Ellefson, G.E. Gibbs, R.W. Baker: J. Vac. Sci. Technol. A **1**, 875 (1983)
3.14 R. Lässer, K.-H Klatt: Phys. Rev. B **28**, 748 (1983)
3.15 D.H.W. Carstens: J. Phys. Chem. **85**, 778 (1981)
3.16 R.H. Steinmeyer, J.D. Braun: In *Radiation Effects and Tritium Technology for Fusion Reactors*, Vol.IV, ed. by J.S. Watson, F.W. Wiffen, USERDA Conf.-750989, Oak Ridge, TN (1986) p.176
3.17 M. Matsuyama, H. Miyake, K. Ashida, K. Watanabe: J. Nucl. Mater. **110**, 296 (1982)
3.18 K. Watanabe, M. Matsuyama, K. Ashida, T. Takeuchi: J. Nucl. Mater. **99**, 320 (1981)
3.19 P.A. Finn, D.K. Sze: Fusion Technology **10**, 1362 (1986)
3.20 J.E. Nasise, R.V. Carlson, R.A. Jalbert: Fusion Technology **10**, 1334 (1986)
3.21 S.R. Bokwa, J.M. Miller, W.J. Holtslander, R.E. Johnson: Nucl. Instr. Methods Phys. Res. A **257**, 52 (1987)
M. Nishikawa, H. Kido, K. Kotoh, M. Sugisaki: J. Nucl. Mater. **115**, 101 (1983)
3.22 H. Weichelgartner: Proc. 12th Symp. on Fusion Technology (SOFT 1982), Vol.1 (Pergamon, London 1983) p.573
H. Weichselgartner: Umschau **13**, 420 (1984)
3.23 V.A. Maroni, E.H. van Deventer: J. Nucl. Mater. **85&86**, 257 (1979)
3.24 C.E. Ells: Atomic Energy of Canada Ltd. Report AECL 6972 (1980)
3.25 J.L. Anderson, D.O. Coffin, J.E. Nasise, R.H. Sherman, R.A. Jalbert: Fusion Technology **8**, 2413 (1985)
3.26 G.A. Morris: UCRL-52262, Lawrence Livermore Laboratory (1977)
J.T. Gill: J. Vac. Sci. Technol. **17**, 645 (1980)
J.T. Gill, W.E. Moddemann, R.E. Ellefson: J. Vac. Sci. Technol. A **1**, 869 (1983)
3.27 W. Lehn: To be published
3.28 R.S. Carlson: In Proc. Int'l Conf. on Radiation Effects and Tritium Technology for Fusion Reactors, Vol.IV, ed. by J.S. Watson, F.W. Wiffen (Gatlinburg, 1975) p.36
3.29 R.C. Bowman, R.S. Carlson, A. Attalla, C.J. Wiedenheft, R.J. De Sando: In Proc. Symp. on Tritium Technology related to Fusion Reactor Systems, ed. by W.H. Smith, W.R. Wilkes, L.J. Wittenberg (1975) p.89
3.30 M.S. Ortman, T.J. Warren, D.J. Smith: Fusion Technology **8**, 2330 (1985)

3.31 D.H.W. Carstens, W.R. David: In Proc. Miami Int'l Symposium on Metal-Hydrogen Systems (1981), ed. by T.N. Veziroglu, p.667
3.32 D.H.W. Carstens: J. Phys. Chem. **85**, 778 (1981)
D.H.W. Carstens, W.R. David: In Proc. Miami Int'l Symp. on Metal-Hydrogen Systems (1981), ed. by T.N. Veziroglu, p.477
3.33 S.A Steward: UCRL-Report-53441, Lawrence Livermore National Laboratory (1983)
3.34 P.C. Souers (ed.): *Hydrogen Properties for Fusion Energy* (Univ. California Press, Berkely 1986)
3.35 H. Ihle, U. Kurz, G. Stöcklin: In Proc. Int'l Conference Radiation Effects and Tritium Technology for Fusion Reactors, Vol.IV, ed. by J.S. Watson, F.W. Wiffen (Gatlinburg 1975) p.414
3.36 Y.I. Zvezdin, Y.I. Belyakov: Soviet Mater. Sci. **3**, 255 (1967)
3.37 G.R. Caskey, Jr., R.G. Derrick: Scr. Metall. **10**, 377 (1976)
3.38 G.R. Caskey, Jr., R.G. Derrick, M.R. Louthan: Scr. Metall. **8**, 481 (1974)
3.39 D.R. Begeal: J. Vac. Sci. Technol. **15**, 1146 (1978)
3.40 O.D. Gonzalez: Trans. Metall. Soc. AIME **239**, 929 (1967)
3.41 A. Van Wieringen, N. Warmoltz: Physica **22**, 849 (1956)
3.42 W.T. Chandler, R.J. Walter: In *Refractory Metal Alloys*, ed. by I. Machlin, R.T. Begley, E.D. Weisert (Plenum, New York 1968) p.197
3.43 Y. Ebisuzaki, W.J. Kass, M. O'Keeffe: J. Chem. Phys. **46**, 1378 (1967)
3.44 S.A. Koffer, J.B. Hudson, G.S. Ansell: Trans. Metall. Soc. AIME **245**, 1735 (1969)
3.45 Y. Ebisuzaki, W.J. Kass, M. O'Keeffe: J. Chem. Phys. **49**, 3329 (1968)
3.46 W.A. Swansiger: In Proc. Int'l Conference on Radiation Effects and Tritium Technology for Fusion Reactors, Vol.IV, ed. by J.S. Watson, F.W. Wiffen (Gatlinburg 1975) p.401
3.47 E. Veleckis, R.K. Edwards: J. Phys. Chem. **73**, 683 (1969)
3.48 J. Völkl, G. Alefeld: In *Diffusion in Solids: Recent Developments*, ed. by A.S. Nowick, J.J. Burton (Academic, New York 1974) p.232
3.49 E.A. Aitken, H.C. Brassfield, P.K. Conn, E.C. Duderstadt, R.E. Fryxell: Trans. Metall. Soc. AIME **239**, 1565 (1967)
3.50 P.C. Souers, I. Moen, R.O. Lindahl, R.T. Tsugawa: J. Am. Ceram. Soc. **61**, 42 (1978)

Chapter 4

4.1 H. Züchner, B. Hüser: In *Gase in Metallen*, ed. by D. Hirschfeld (Deutsche Gesellschaft für Metallkunde, E.V. 1984) p.53, ISBN 3-88355-086-8
H. Züchner, B. Hüser, P. Kock: Z. Analytische Chem. **329**, 169 (1987)
B. Ladna, C.M. Loxton, H.K. Birnbaum: Acta Metall. **34**, 899 (1986)
4.2 M.E. Malinowski: J. Vacuum Sci: Technol. **15**, 39 (1978)
B.C. Lamartine, T.W. Haas, J.S. Solomon: Appl. Surf. Sci. **4**, 537 (1980)
M.E. Malinowski: J. Vacuum Sci. Technol. **19**, 120 (1981)
P. Bracconi, R. Lässer: Appl. Surf. Sci. **28**, 204 (1987)
P. Bracconi, E. Pörschke, R. Lässer: Appl. Surf. Sci. **32**, 392 (1988)
4.3 R.M. Cotts: In *Hydrogen in Metals I*, ed. by G. Alefeld, J. Völkl, Topics Appl. Phys. Vol.28 (Springer, Berlin, Heidelberg 1978) p.227
R.C. Bowman, Jr.: Nature **271**, 531 (1978)
P.C. Souers, T.A. Jolly, C.F. Cline: J. Phys. Chem. Solids **28**, 1717 (1967)

4.4 H. Peisl: In *Hydrogen in Metals*, ed. by G. Alefeld, J. Völkl, Topics Appl. Phys. Vol.28 (Springer, Berlin, Heidelberg 1978) p.52 and references therein
4.5 R. Feenstra: Dissertation, Vrije Universiteit, Amsterdam (1985)
R. Feenstra, R. Griessen, D.G. de Groot: J. Phys. F **16**, 1933 (1986)
R. Feenstra, D.G. de Groot, J.H. Rector, E. Salomons, R. Griessen: J. Phys. F **16**, 1953 (1986)
4.6 S. Okuda, R. Taniguchi, M. Fujishiro, Y. Satoh, E. Hiraoka; J. Appl. Phys. **54**, 6790 (1983): ibid **128–129**, 725 (1984)
J.A. Sawicki: J. Nucl. Mater. **141–143**, 327 (1986); Nucl. Instru. Meth. Phys. Res. B **23**, 521 (1987)
L.G. Earwaker, J.B.A. England, D.J. Goldie: Nucl. Instru. Meth. in Physics Research **B24/25**, 711 (1987)
4.7 T. Springer: *Quasielestic Neutron Scattering for the Investigation of Diffuse Motions in Solid and Liquids*, Springer Tracts Mod. Phys., Vol.64 (Springer, Berlin, Heidelberg 1972)
T. Springer: In *Hydrogen in Metals I*, ed. by G. Alefeld, J. Völkl, Topics Appl. Phys. Vol.28 (Springer, Berlin, Heidelberg 1978) p.75
D.G. Hunt, D.K. Ross: J. Less Common Metals **49**, 169 (1976)
D. Richter: In *Exotic Atoms '79*, ed. by K. Crowe, J. Duclos, G. Fiorentini, G. Torelli (Plenum, New York 1980) p.245
R. Hempelmann: J. Less Common Metals **101**, 69 (1984)
4.8 H. Rauch, A. Zeilinger: Atomic Energy Rev. **15**, 249 (1977) (IAEA Vienna 1977)
4.9 C.A. Colmenares: Nucl. Instrum. Meth. **114**, 269 (1974)
M. Sugisaki, K. Idemitsu, F. Furuya: Radiochem. Radioanal. Lett. **51**, 293 (1982)
G.F. Knoll: *Radiation Detection and Measurement* (Wiley, London 1979)
T. Aoyama, H. Miyai, T. Watanabe: Nucl. Instr. Meth. Phys. Res. **221**, 644 (1984)
K.Y. Wong, R.G.C. McElroy: Fusion Technology **8**, 2082 (1985)
M. Matsuyama, K. Ichimura, K. Ashida, K. Watanabe, H. Sato: Fusion Technology **8**, 2461 (1985)
4.10 G.R. Caskey, Jr.: In *Advanced Techniques for Characterizing Hydrogen in Metals*, ed. by N.F. Fiore, B.J. Berkowitz (Society of AIME, Louisville, Kentucky 1981)
4.11 J.W. Pyper, E.M. Kelly, J.G. Magistad, R.T. Tsugawa, P.E. Roberts, P.C. Souers: UCRL-52391, Lawrence Livermore Laboratory (1978)
R.E. Ellefson, W.E. Moddeman, H.F. Dylla: J. Vac. Sci. Technol. **18**, 1062 (1981)
J. von Seggern, S. Berger, M. Erdweg, W.O. Hofer: J. Vac. Sci. Technol. A 2, 1516 (1984)
4.12 M. Saeki, T. Hirabayashi, Y. Aratono, T. Hasegawa, E. Tachikawa: J. Nucl. Sci. Technol. **20**, 762 (1983)
J. King: J. Phys. Chem. **67**, 1397 (1963)
C. Genty, R. Schott: Analytical Chemistry 42, 7 (1970)
M.L. Conti, M. Lesimple: J. Chromatogr. 29, 32 (1967)
T. Schober, C. Dieker: Rev. Sci. Instrum. **58**, 1116 (1987)
4.13 R.E. Setchell, D.K. Otteson: Report SAND 74-8644, Sandia Laboratories, Livermore, CA 94550 (1975)
P.C. Souers (ed.): *Hydrogen Properties for Fusion Energy* (Univ. California, Berkeley 1986)
4.14 R. Lässer, K. Bickmann: J. Nucl. Mater. **126**, 234 (1984)

4.15 C.W. Pennington, T.S. Elleman, K. Verghese: Nuclear Technology **22**, 405 (1974)
T.S. Ellemann, K. Verghese: J. Nucl. Materials **53**, 299 (1974)
R. Lässer, K. Bickmann: J. Nucl. Mater. **132**, 244 (1985)
4.16 R. Lässer, H. Wenzl: In *Proc.12. Symposium on Fusion Technology 1982*, Vol.II (Pergamon, Oxford 1983) p.783
4.17 R. Lässer: Z. Phys. Chem. N.F. **143**, 23 (1985)
4.18 V.A. Lubimov, E.G. Novikov, V.Z. Nozik, E.F. Tretyakov, V.S. Kosik: Phys. Lett. **94B**, 266 (1980)
I.G. Kaplan, V.N. Smutnyi, G.V. Smelov: Sov. Phys. JETP **57**, 483 (1983)
T. Kirsten, H. Richter, E. Jessberger: Phys. Rev. Lett. **50**, 474 (1983)
W. Kündig: Spektrum der Wissenschaft (1986) p.12; Phys. Rev. B **42**, 380 (1986)
G.L. Borchert, D. Gotta, H.R. Koch, R. Salziger, H.J. Probst, O.W.B. Schult, R. Lässer: to be published
E. Amaldi: Phys. Rep. **111**, 1 (1984)
4.19 J.E. Pearson: Applied Spectroscopy **27**, 450 (1973)
4.20 M.E. Malinowski: Appl. Phys. Lett. **39**, 509 (1981)
4.21 M.E. Malinowski: J. Vac. Sci. Technol. **A1**, 933 (1983)
4.22 M.E. Malinowski: J. Less-Common Metals **89**, 1 (1983)
4.23 A.W. Rogers: *Techniques of Autoradiography* (Elsevier, Amsterdam 1979)
4.24 H.A. Fischer, G. Werner: *Autoradiography* (de Gruyter, Berlin 1971)
4.25 T. Iida, Y. Ikebe: Nucl. Instrum. Meth. Phys. Res. A **253**, 119 (1986)
4.26 R.H. Herz: *The Photographic Action of Ionizing Radiations* (Wiley, New York 1969)
4.27 M. Aucouturier, G. Lapasset, T. Asaoka: Metallography **11**, 5 (1978)
4.28 I. Taguchi: Proc. Hydrogen in Metals (JIMIS-2) p.225
4.29 C.R. Cupp, P. Flubacher: J. Nucl. Mater. **6**, 213 (1962)

Chapter 5

5.1 E. Wicke, H. Brodowsky, H. Züchner: In *Hydrogen in Metals II*, ed. by G. Alefeld, J. Völkl, Topics Appl. Phys. Vol.29 (Springer, Berlin, Heidelberg 1978) p.73
5.2 F.A. Lewis: Platinum Met Rev. **26**, 20 (1982); **26**, 70 (1982); **26**, 121 (1982)
F.A. Lewis: *The Palladium Hydrogen System* (Academic, New York 1967)
5.3 T.B. Flanagan, W.A. Oates: Adv. Chem. **167**, 283 (1978)
5.4 A. Sieverts, G. Zapf: Z. Phys. Chem. **174A**, 359 (1935)
5.5 A. Sieverts, W. Danz: Z. Phys. Chem. **38B**, 46 (1937)
5.6 L.J. Gillespie, W.R. Downs: J. Am. Chem. Soc. **61**, 2496 (1939)
5.7 E. Wicke, G.H. Nernst: Ber. Bunsenges. Phys. Chem. **68**, 224 (1964)
5.8 H. Brodowsky: Z. Phys. Chem. **44**, 129 (1965)
5.9 H. Frieske, E. Wicke: Ber. Bunsenges. Phys. Chem. **77**, 48 (1973)
5.10 J.D. Clewley, T. Curran, T.B. Flanagan, W.A. Oates: J. Chem. Soc. Faraday Trans. 1 **69**, 449 (1973)
5.11 W.A. Oates, T.B. Flanagan: J. Chem. Soc. Faraday Trans. 1 **73**, 407 (1977)
5.12 G. Boureau, O.J. Kleppa, P. Dantzer: J. Chem. Phys. **64**, 5247 (1976)
5.13 G. Boureau, O.J. Kleppa: J. Chem. Phys. **65**, 3915 (1976)
5.14 C. Picard, O.J. Kleppa, G. Boureau: J. Chem. Phys. **69**, 5549 (1978)
5.15 C. Labes, R.B. McLellan: Acta Metall. **26**, 893 (1978)
5.16 T. Kuji, W.A. Oates, B.S. Bowerman, T.B. Flanagan: J. Phys. F **13**, 1785 (1983)

5.17 E. Wicke, J. Blaurock: Ber. Bunsenges. Phys. Chem. **85**, 1091 (1981)
5.18 R. Feenstra, R. Griessen, D.G. de Groot: J. Phys. F **16**, 1933 (1986)
5.19 R.L. Favreau, R.E. Patterson, D. Randall, O.N. Salmon: Knolls Atomic Power Laboratory, U.S. AEC Report No. KAPL-1036, 1954
5.20 S. Schmidt, G. Sicking: Z. Naturforsch. **33a**, 1328 (1978)
5.21 R. Lässer: Phys. Rev. **B26**, 3517 (1982)
5.22 R. Lässer, K.-H. Klatt: Phys. Rev. **B28**, 748 (1983)
5.23 R. Lässer: Phys. Rev. **B29**, 4765 (1984)
5.24 R. Lässer: J. Phys. F **14**, 1975 (1984)
5.25 R.C. Bowman, R.S. Carlson, A. Attalla, C.J. Wiedenheft, R.J. DeSando: In Proc. Symp. on Tritium Technology Related to Fusion Reactor Systems, ed. by W.H. Smith, W.R. Wilkes, L.J. Wittenberg (1975) p.89
5.26 G. Sicking: J. Less-Common Metals **101**, 169 (1984)
5.27 R. Lässer: In *Nichtmetalle in Metallen*, ed. by D. Hirschfeld (Deutsche Gesellschaft für Metalkunde e.V. 1987) p.25
5.28 R. Lässer, G.L. Powell: Phys. Rev. **B34**, 578 (1986)
5.29 G.L. Powell: J. Phys. Chem. **80**, 375 (1976)
5.30 J.J. Rush, J.M. Rowe, D. Richter: Z. Phys. **B55**, 283 (1984)
5.31 Y. Ebisuzaki, M. O'Keeffe: In, *Progress in Solid State Chemistry*, Vol.4, ed. by H. Reiss (Pergamon, New York 1967) p.187
5.32 Y. Ebisuzaki, W.J. Kass, M. O'Keeffe: J. Chem. Phys. **46**, 1373 (1967)
5.33 Y. Ebisuzaki, W.J. Kass, M. O'Keeffe: J. Chem. Phys. **49**, 3329 (1968)
5.34 R. Hempelmann, D. Richter, D.L. Price: Phys. Rev. Lett. **58**, 1016 (1987)
5.35 R. Lässer: J. Less-Common Metals **131**, 263 (1987)
5.36 T.B. Flanagan, T. Kuji, W.A. Oates: J. Phys. F **15**, 2273 (1985)
5.37 W.A. Oates, R. Lässer, T. Kuji, T.B. Flanagan: J. Phys. Chem. Solids **47**, 429 (1986)
5.38 B.M. Geerken, R. Griessen, L.M. Huisman, E. Walker: Phys. Rev. **B26**, 1637 (1982)
5.39 R. Lässer: Z. Phys. Chem. NF **143**, 23 (1985)
5.40 E. Wicke: Z. Phys. Chem. NF **143**, 1 (1985)
5.41 H. Brodowsky, E. Pöschel: Z. Phys. Chem. **44**, 143 (1965)
5.42 H. Buck: JÜL-Report-722FF, Kernforschungsanlage Jülich (1970)
5.43 G.L. Holleck: J. Phys. Chem. **74**, 503 (1970)
5.44 P.T. Gallagher, W.A. Oates: J. Phys. Chem. Solids **32**, 2105 (1971)
5.45 H. Buck, G. Alefeld: Phys. Status Solidi **B49**, 317 (1972)
5.46 G. Boureau, O.J. Kleppa, K.C. Hong: J. Chem. Phys. **67**, 3437 (1977)
5.47 G. Boureau, O.J. Kleppa; P.D. Antoniou: J. Solid State Chem. **28**, 223 (1979)
5.48 C. Picard, O.J. Kleppa, G. Boureau: J. Chem. Phys. **70**, 2710 (1979)
5.49 W. Loss: Dissertation (Universität München 1976)
5.50 J. Blaurock, Dissertation (Universtät Münster 1985)
5.51 M. Yoshihara, R.B. McLellan: Acta Metall. **33**, 83 (1985)
5.52 R. Lässer, G.L. Powell: J. Less-Common Metals **130**, 387 (1987)
5.53 M.R. Chowdhury, D.K. Ross: Solid State Commun. **13**, 229 (1973)
5.54 P. Fratzl, O. Blaschko, E. Walker: Phys. Rev. **B34**, 164 (1986)
5.55 R. Lässer, G.L. Powell: Fusion Technology **14**, 695 (1988)
R. Lässer, P. Meuffels, R. Feenstra: JÜL-Report 2183, Kernforschungsanlage Jülich (1988) ISSN 0366-0885
5.56 W.A. Oates, R. Ramanathan: In Proc. 2nd Int'l Cong. on Hydrogen in Metals, Paris, France (1977) Vol.3, 2A11, p.1-8
5.57 R. Griessen: In Hydrogen in Disordered and Amorphous Solids, ed. by G. Bambakidis, R.C. Bowman, Jr., NATO ASI Series **136**, 153 (1986)

5.58 E. Veleckis, J. Nucl. Mater. **79**, 20 (1979)
5.59 G.M. Begun, J.F. Land, J.T. Bell: J. Chem. Phys. 72, 2959 (1980)
5.60 J.J. Rush, H.E. Flotow, D.W. Connor, C.L. Thaper: J. Chem Phys. **45**, 3817 (1966)
5.61 H.P. Bleichert: JÜL-Report 2005, Kernforschungsanlage Jülich (1985) ISSN 0366-0885
5.62 P. Meuffels: JÜL-Report 2081, Kernforschungsanlage Jülich (1986) ISSN 0366-0885
P. Meuffels, R. Lässer, H. Wenzl: to be published
5.63 E. Veleckis, R.K. Edwards: J. Phys. Chem. 73, 683 (1969)
5.64 O.J. Kleppa, P. Dantzer, M.E. Melchinak: J. Chem. Phys. **61**, 4048 (1974)
5.65 D. Richter: private communication
5.66 G. Alefeld, J. Völkl (eds.): *Hydrogen in Metals I,II*, Topics Appl. Phys. Vols.28,29 (Springer, Berlin, Heidelberg 1978)
E. Fromm, E. Gebhardt: *Gase und Kohlenstoff in Metallen* (Springer, Berlin, Heidelberg 1976)
J.W. Ward: In *Handbook on the Physics and Chemistry of the Actinides*, ed. by A.J. Freeman, C. Keller (Elsevier, Amsterdam 1985) Chap.1
5.67 V.A. Somenkov, S.Sh. Shil'shtein: Progress in Mater. Sci. **24**, 267 (1980)
5.68 T. Schober, H. Wenzl: In *Hydrogen in Metals II*, ed. by G. Alefeld, J. Völkl, Topics Appl. Phys. Vol.29 (Springer, Berlin, Heidelberg 1978) p.11
5.69 G. Alefeld: Phys. Status Solidi **32**, 67 (1969)
5.70 R. Lässer: J. Phys. Chem. Solids **46**, 33 (1985)
5.71 R. Lässer, G.L. Powell: In *Hydrogen in Disordered and Amorphous Solids*, ed. by G. Bambakidis, R.C. Bowman, Jr. (Plenum, New York 1986) p.387
5.72 R. Lässer, K. Bickmann: J. Nucl. Mater. **126**, 234 (1984)
5.73 R.C. Bowman, Jr., A. Attalla, W.E. Tadlock, D.E. Sullenger, R.L. Yauger: Scripta Metall. **16**, 933 (1982)
5.74 Zh. Qi, J. Völkl, R. Lässer, H. Wenzl: J. Phys. F **13**, 2053 (1983)
5.75 R. Lässer, T. Schober: J. Less-Common Met. **130**, 453 (1987)
5.76 R. Lässer, K. Bickmann: J. Nucl. Mater. **132**, 244 (1985)
5.77 O. Blaschko, R. Klemencic, P. Weinzierl, O.J. Eder, P. von Blanckenhagen: Acta Crystallogr. A**36**, 605 (1980)
5.78 O. Blaschko, P. Fratzl, R. Klemencic: Phys. Rev. B**24**, 6486 (1981)
5.79 O. Blaschko: J. Less-Common Metals **100**, 307 (1984)
5.80 R. Lässer, K.-H. Klatt, P. Mecking, H. Wenzl: JÜL-Report 1800, Kernforschungsanlage Jülich (1982) ISSN 0366-0885
5.81 J.F. Smith, D.T. Peterson: Bulletin of Alloy Phase Diagrams **3**, 49 (1982), and **3**, 55 (1982)
5.82 T. Schober: In *Electronic Structure and Properties of Hydrogen in Metals*, ed. by P. Jena, C.B. Satterthwaite (Plenum, New York 1983) p.1
H. Wenzl, P. Meching, M.E. de Prieto: Ann. Chim. Fr. **7**, 517 (1982)
5.83 G. Alefeld, J. Völkl, G. Schaumann: Phys. Status Solidi **37**, 337 (1970)
5.84 J. Völkl, G. Alefeld: Z. Phys. Chem. NF **114**, 123 (1979)
5.85 R. Heller, H. Wipf: Phys. Status Solidi A**33**, 525 (1976)
5.86 T. Schober, A. Carl: Phys. Status Solidi A**43**, 443 (1977)
5.87 T. Schober: Scripta Metall. **12**, 549 (1978)
5.88 H. Peisl: In *Hydrogen in Metals I*, ed. by G. Alefeld, J. Völkl, Topics Appl. Phys. Vol.28 (Springer, Berlin, Heidelberg 1978) p.53
5.89 T.B. Flanagan, T. Schober, H. Wenzl: Acta Metall. **33**, 685 (1985)
5.90 C.K. Hall: In *Electronic Structure and Properties of Hydrogen in Metals*, ed. by P. Jena, C.B. Satterthwaite (Plenum, New York 1983) p.11

C.K. Hall, M. Futran: J. Less-Common Metals 74, 237 (1980)
5.91 T.B. Flanagan, W.A. Oates, S. Kiaashimoto: Acta Metall. 31, 199 (1983)
5.92 I.R. Entin, V.A. Somenkov, S.Sh. Shil'shtein: Sov. Phys. Solid St. 16, 1569 (1975)
5.93 J.F. Smith: Bull. Alloy Phase Diagrams 4, 39 (1983)
5.94 H. Wenzl, J.-M. Welter: In *Current Topics in Material Science I*, ed. by E.Kaldis (1978) p.603
5.95 H. Zabel: Dissertation (Universität München 1978)
5.96 H. Zabel, H. Peisl: J. Phys. F 9, 1461 (1979)
5.97 J.-M. Welter, M.A. Pick, T. Schober, J. Hauck, H.J. Fenzl, H. Wenzl: In Proc. 2nd Int'l Congr. on H in Metals, Paris (1977) 1D3
5.98 K. Fujita, Y.C. Huang, M. Tada: J. Jpn. Inst. Met. 43, 601 (1979)
5.99 T. Kuji, T.B. Flanagan: J. Phys. F 15, L59 (1985)
5.100 U. Köbler, J.-M. Welter: J. Less-Common Metals 84, 255 (1982)
5.101 W. Zag: Dissertation, Max-Planck-Institut für Metallforschung, Institute für Physik, Stuttgart (1985)
5.102 A. Craft, T. Kuji, T.B. Flanagan: J. Phys. F 18, 1149 (1988)
5.103 D. Richter, S.M. Shapiro: Phys. Rev. B22, 599 (1980)
5.104 R. Hempelmann, D. Richter, A. Kollmar: Z. Phys. B44, 159 (1981)
5.105 J. Eckert, J.A. Goldstone, D. Tonks, D. Richter: Phys. Rev. B 27, 1980 (1983)
5.106 J.J. Rush, A. Magerl, J.M. Rowe, J.M. Harris, J.L. Provo: Phys. Rev. B 24, 4903 (1981)
5.107 C.W. Pennington, T.S. Elleman, K. Verghese, Nuclear Technology 22, 405 (1974)
D. Chandra, T.S. Elleman, K. Verghese: J. Nuclear Mater. 59, 263 (1976)
5.108 C.B. Satterthwaite, I.L. Toepke: Phys. Rev. Lett. 25, 741 (1970)
5.109 T. Skoskiewicz: Phys. Status Solidi A 11, K 123 (1972)
5.110 B. Stritzker, W. Buckel: Z. Physik 257, 1 (1972)
5.111 J.E. Schirber, J.M. Mintz, W. Wall: Solid State Commun. 52, 837 (1984)
5.112 J.E. Schirber, C.J.M. Northrup, Jr.: Phys. Rev. B 10, 3828 (1974)
5.113 W.L. McMillan: Phys. Rev. 167, 331 (1968)
5.114 B.N. Ganguly: Z. Phys. 265, 433 (1973)
B.N. Ganguly: Phys. Rev. B 14, 3848 (1976)
5.115 R.J. Miller, C.B. Satterthwaite: Phys. Rev. Lett 34, 144 (1975)
5.116 D.A. Papaconstantopoulos, B.M. Klein, E.N. Economou, L.L. Boyer: Phys. Rev. B 17, 141 (1978)
5.117 P.B. Allen, R.C. Dynes: Phys. Rev. B 12, 905 (1975)
5.118 J.M. Rowe, J.J. Rush, H.G. Smith, M. Mostoller, H.E. Flotow: Phys. Rev. Lett. 33, 1297 (1974)
5.119 A. Rahman, K. Sköld, C. Pelizzari, S.K. Sinha, H. Flotow: Phys. Rev. B 14, 3630 (1976)
5.120 R. Griessen, D.G. de Groot: Helv. Phys. Acta 55, 699 (1982)
5.121 P. Jena, J. Jones, R.M. Nieminen: Phys. Rev. B 29, 4140 (1984)
5.122 A. Magerl, H. Wipf, G. Alefeld: J. Phys. Chem. Solids 38, 683 (1977)
5.123 T. Springer: In *Hydrogen in Metals I*, ed. by G. Alefeld, J. Völkl, Topics Appl. Phys. Vol.28 (Springer, Berlin, Heidelberg 1978) p.75
5.124 J.M. Rowe, J.J. Rush, J.E. Schirber, J.M. Mintz: Phys. Rev. Lett. 57, 2955 (1986)
5.125 H.L.M. Bakker, M. van Sprang, R. Griessen: J. Phys. F 15, 63 (1985)
5.126 H.L.M. Bakker: Dissertation (Vrije Universiteit, Amsterdam 1985)
5.127 H.L.M. Bakker, R. Griessen, N.J. Koeman, R. Albers, G.H. Sicking: J. Phys. F 16, 721 (1986)

5.128 E. Wicke: J. Less-Common Metals **101**, 17 (1984)
5.129 C. Wert, C. Zener: Phys. Rev. **76**, 1169 (1949)
5.130 G.H. Vineyard: J. Phys. Chem. Solids **3**, 121 (1957)
5.131 A.D. Le Claire: Philos. Mag. **14**, 1271 (1966)
5.132 Y. Ebisuzaki, W.J. Kass, M. O'Keeffe: Philos. Mag. **15**, 1071 (1967)
5.133 L. Katz, M. Guinan, R.J. Borg: Phys. Rev. B **4**, 330 (1971)
5.134 K.W. Kehr: In *Hydrogen in Metals*, ed. by G. Alefeld, J. Völkl, Topics Appl. Phys. Vol.28 (Springer, Berlin, Heidelberg 1978) p.197
K.W. Kehr: In Electronic Structure and Properties of Hydrogen in Metals, ed. by P. Jena, C.B. Satterthwaite, NATO Conference Series, Series VI, 1983, p.531
D. Richter: In *Exotic Atoms '79*, ed. by K. Crowe, J. Duclos, G. Fiorentini, G. Torelli (Plenum, London 1980) p.245
5.135 C.P. Flynn, A.M. Stoneham: Phys. Rev. B **1**, 3966 (1970)
5.136 A.M. Stoneham: In *Exotic Atoms 79*, ed. by K. Krowe, J. Nuclos, G. Fiorentini, G. Torelli (Plenum, New York 1980) p.209
5.137 Yu Kagan, M.I. Klinger: J. Phys. C **7**, 2791 (1974)
5.138 D. Emin, M.I. Baskes, W.D. Wilson: Phys. Rev. Lett. **42**, 791 (1979)
Z. Phys. Chem. NF **114** 231 (1979)
5.139 M. Kokkinidis: Dissertation (Technische Universität, München 1977)
H. Rückel: Dissertation (Technische Universität, München 1979)
5.140 G. Schaumann, J. Völkl, G. Alefeld: Phys. Rev. Lett. **21**, 891 (1968)
Phys. Status Solidi **42**, 401 (1970)
5.141 U. Freudenberg, J. Völkl, J. Bressers, G. Alefeld: Scripta Metall. **12**, 165 (1978)
5.142 G. Matusiewicz, H.K. Birnbaum: J. Phys. F **7**, 2285 (1977)
5.143 G. Sicking, H. Buchold: Z. Naturforsch. A **26**, 1973 (1971)
5.144 M. Sugisaki, H. Furuya, S. Mukai: J. Less-Common Metals **107**, 79 (1985)
5.145 R. Messer, D. Höpfel, C. Schmidt, A. Seeger, W. Zag, R. Lässer: Z. Phys. Chem. NF **145**, 179 (1985)
5.146 R. Messer, A. Blessing, S. Dais, D. Höpfel, G. Majer, C. Schmidt, A. Seeger, W. Zag, R. Lässer: Z. Phys. Chem. NF Suppl.-H2, 61 (1986)
5.147 H. Teichler, A. Klamt: Phys. Lett. A **108**, 281 (1985)
H.R. Schober, A.M. Stoneham: Phys. Rev. Lett. **60**, 2307 (1988)
5.148 R.C. Bowman, Jr., A. Attalla, B.D. Craft: Scripta Metall. **17**, 937 (1983)
5.149 R.C. Bowman, Jr., A. Attalla, B.D. Craft: Fusion Technology **8**, 2366 (1985)
5.150 G. Sicking, M. Glugla, H. Huber: Ber. Bunsenges. Phys. Chem. **87**, 418 (1983)
5.151 H. Kronmüller: In *Hydrogen in Metals I*, ed. by G. Alefeld, J. Völkl, Topics Appl. Phys. Vol.28 (Springer, Berlin, Heidelberg 1978) p.289
5.152 H. Kronmüller: *Nachwirkung in Ferromagnetika* (Springer, Berlin, Heidelberg 1968) p.181
5.153 B. Hohler, H. Kronmüller: Philos. Mag. **43**, 1189 (1981)
5.154 H. Kronmüller, G. Higelin, P. Varges, R. Lässer: Z. Phys. Chem. NF **143**, 143 (1985)
5.155 G. Higelin, H. Kronmüller, R. Lässer: Phys. Rev. Lett. **53**, 2117 (1984)
5.156 R.R. Arons, H.G. Bohn, H. Lütgemeier: Solid State Commun. **14**, 1203 (1974)
5.157 G. Alefeld, J. Völkl: In *Hydrogen in Metals I* ed. by G. Alefeld, J. Völkl, Topics Appl. Phys. Vol.28 (Springer, Berlin, Heidelberg 1978) p.321
5.158 L. Katz, M. Guinan, R.J. Borg: Phys. Rev. B **4**, 330 (1971)
5.159 B. Hohler, H. Schreyer: J. Phys. F **12**, 857 (1982)
5.160 J. Hauck: Z. Physik. Chemie NF **114**, 165 (1979)
5.161 S.M. Fries, H.G. Wagner, S.J. Campbell, U. Gonser, N. Blaes, P. Steiner: J. Phys. F **15**, 1179 (1985)
A.J. Maeland: In *Hydrogen in Disordered and Amorphous Solids*, ed. by G.

Bambakidis, R.C. Bowman, Jr., NATO ASI Series, Physics, Vol.136 (Plenum, New York 1986) p.127
5.162 T. Karasinski, K. Patzelt, C. Dieker, H. Hansen, H. Wenzl, T. Schober: Kernforschungsanlage Jülich JÜL-Report-2136 1987 ISSN 0366-0885
T. Schober, C. Dieker, R. Feenstra: J. Phys. F **18**, 1119 (1988)
5.163 H. Pfeiffer, H. Peisl: Phys. Lett **60A** 363 (1977)
5.164 A. Magerl, B. Berre, G. Alefeld: Phys. Status Solidi A **36**, 161 (1976)
5.165 D.T. Peterson, H.M. Herro: Met. Trans. **14A**, 17 (1983)

Chapter 6

6.1 J.F. Remark, A.B. Johnson, Jr., H. Farrar, D.G. Atteridge: Nuclear Technology **29**, 369 (1976)
6.2 P. Jung: Radiation Effects **35**, 155 (1978)
6.3 J. Laakmann: Jül-Report 2007, Kernforschungsanlage Jülich (1985), ISSN 0366-0885
P. Jung, J. Laakmann, H.J. Odenthal: In *Nichtmetalle in Metallen* ed. by D. Hirschfeld (Deutsche Gesellschaft für Metallkunde e.V. 1987) p.101
6.4 H.J. von den Driesch: Jül-Report 1663, Kernforschungsanlage Jülich (1980), ISSN 0366-0885
6.5 H.J. von den Driesch, P. Jung: High temperatures - high pressures **12**, 635 (1980)
6.6 H. Schröder, W. Kesternich, H. Ullmaier: Nuclear Engineering and Design/Fusion **2**, 65 (1985)
6.7 H. Trinkaus: Rad. Eff. **78**, 189 (1983)
6.8 W.D. Wilson, C.L. Bisson, M.I. Baskes: SAND 81-8727 (Sandia Laboratories)
C.L. Bisson; W.D. Wilson: SAND 80-8642 (Sandia Laboratories)
6.9 W.D. Wilson, C.L. Bisson, M.I. Baskes: Phys. Rev. B **24**, 5616 (1981)
6.10 M.I. Baskes, W.D. Wilson: Phys. Rev. B **27**, 2210 (1983)
6.11 D.J. Reed: Radiation Effects **31**, 129 (1977)
6.12 V. Sciani, P. Jung: Radiation Effects **78**, 87 (1983)
6.13 H. Trinkaus: J. Nucl. Mater. **133 + 134**, 105 (1985)
6.14 L.K. Mansur, E.H. Lee, P.J. Maziesz, A.P. Rowcliffee: J. Nucl. Mater. **141-143**, 633 (1986)
6.15 L.K. Mansur, W.A. Coghlan: J. Nucl. Mater. **119**, 1 (1983)
6.16 Proc. Int'l Symp. on Fundamental Aspects of Helium in Metals, ed. by H. Ullmaier. Rad. Eff. **78**, 1-426 (1983)
N.H. Ghoniem, P. Maziasz: PPG-698 (Cter. for Plasma Phys. and Fusion Eng., Univ. California, Los Angeles 1983) and references therein
6.17 A.M. Rodin, V.V. Surenyants: Fiz. Khim. **45**, 1094 (1971)
6.18 M.E. Malinowski, P.R. Coronado: In Radiation Effects and Tritium Technology for Fusion Reactors, Vol.IV, USERDA Conf.-750989, ed. by J.S. Watson, F.W. Wiffen (Oak Ridge, TN 1976) p.53
6.19 W.G. Perkins, W.J. Kass, L.C. Beavis: In Radiation Effects and Tritium Technology for Fusion Reactors, Vol.IV, USERDA Conf.-750989, ed. by J.S. Watson, F.W. Wiffen (Oak Ridge, TN 1976) p.83
6.20 R.C. Bowman, Jr., A. Attalla: Phys. Rev. B **16**, 1828 (1977)
6.21 D.J. Mitchell, R.C. Patrick: J. Vac. Sci. Technol. **19**, 236 (1981)
6.22 D.J. Mitchell, J.L. Provo: J. Appl. Phys. **57**, 1855 (1985)
6.23 R.G. Spulak, Jr.: J. Less-Common Met. **132**, L17 (1987)

6.24 G.J. Thomas, W.A. Swansiger, M.I. Baskes: J. Appl. Phys. **50**, 6942 (1979)
G.J. Thomas: Rad. Eff. **78**, 37 (1983)
6.25 J.R. Cost, R.G. Hickmann: J. Vac. Sci. Technol. **12**, 516 (1975)
6.26 R.C. Bowman, Jr., R.S. Carlson, R.J. DeSando: In Proc. 24th Conf. on Remote Systems Technology (1976) p.62
6.27 R.C. Bowman, Jr., A. Attalla, B.D. Craft: Fusion Technology **8**, 2366 (1985)
6.28 A. Attalla, R.C. Bowman, Jr.: Analytical Chemistry **47**, 728 (1975)
6.29 R.C. Bowman, Jr., A. Attalla: In Radiation Effects and Tritium Technology for Fusion Reactors, Vol.IV, USERDA Conf.-750989, ed by J.W. Watson, F.W. Wiffen (Oak Ridge, TN 1976) p.68
A. Atalla, R.C. Bowman, Jr.: In Tritium Technology in Fission, Fusion, and Isotopic Applications. ANS Meeting, Dayton, OH (1980) p.108
6.30 R.C. Bowman, Jr., R.H. Steinmeyer, L.K. Matson, A. Attalla, B.C. Craft: Fusion Technology **8**, 2337 (1985)
6.31 R.C. Bowman, Jr.: Nature **271**, 531 (1978)
6.32 J.H. Evans, A. van Veen, L.M. Caspers: Scripta Metall. **15**, 323 (1981)
6.33 G.J. Thomas, R. Bastasz: J. Appl. Phys. **52**, 6426 (1981)
6.34 G.J. Thomas, J.M. Mintz: J. Nucl. Mater. **116**, 336 (1983)
6.35 W. Jäger, R. Lässer, T. Schober, G.J. Thomas: Rad. Eff. **78**, 165 (1983)
6.36 T. Schober, R. Lässer, W. Jäger, G.J. Thomas: J. Nucl. Mater. **122&123**, 571 (1984)
6.37 T. Schober, G.J. Thomas, R. Lässer, W. Jäger: Scripta Metall. **18**, 255 (1984)
6.38 T. Schober, R. Lässer: J. Nucl. Mater. **120**, 137 (1984)
6.39 T. Schober, H. Trinkaus, R. Lässer: J. Nucl. Mater. **141-143**, 453 (1986)
6.40 N.L. Barmore, C.J. Echer, E.L. Raymond, R.R. Vandervoort: Metallurgical Trans. **11 A**, 239 (1980)
G.R. Caskey, Jr., D.E. Rawl, Jr., D.A. Mezzanotte, Jr.: Scripta Metall. **16**, 969 (1982)
G.R. Caskey, Jr.: Fusion Technology **8**, 2293 (1985)
M.R. Louthan, Jr., G.R. Caskey, Jr., D.E. Rawl, Jr. C.W. Krapp: In Radiation Effects and Tritium Technology for Fusion Reactors, Vol.IV, USERDA Conf-750989, ed. by J.S. Watson, F.W. Wiffen (Oak Ridge, TN 1976) p.98
6.41 G.C. Abell, A. Attalla: Phys. Rev. Lett. **59**, 995 (1987)
6.42 R. Kirchheim: Acta Metall. **29**, 835 (1981); **29**, 845 (1981); **30**, 1069 (1982)
6.43 H.N.G. Wadley, R. Mehrabian: Mater. Sci. Eng. **65**, 245 (1984)
6.44 G. Canelli, R. Cantelli: Mater. Sci. Eng. **48**, 107 (1981)
6.45 T. Schober, J. Golczewski, R. Lässer, C. Dieker, H. Trinkaus: Z. Phys. Chem. NF **147**, 161 (1986)
6.46 J. Golczewski, T. Schober: Mater. Sci. Eng. **68**, L5 (1984)
6.47 P.M.S. Jones, W. Edmondson, N.J. McKenna: J. Nucl. Mater. **23**, 309 (1967)
6.48 L.C. Beavis, C.J. Miglionico: J. Less-Common Metals **27**, 201 (1972)
6.49 T. Schober, R. Lässer, J. Golczewski, C. Dieker, H. Trinkaus: Phys. Rev. **B31**, 7109 (1985)
6.50 T. Schober: J. Phys. E **17**, 196 (1984)
6.51 T. Schober, R. Lässer, C. Dieker, H. Trinkaus: J. Less-Common Metals **131**, 293 (1987)
6.52 R. Manzke, M. Campagna: Solid Sate Commun. **39**, 313 (1981)
R. Manzke, G. Crecelius, J. Fink, H. Trinkaus, W. Jäger: J. Phys. F **12**, L279 (1982)
6.53 W. Jäger, R. Manzke, H. Trinkaus, R. Zeller, J. Fink, G. Crecelius: Rad. Eff. **78**, 315 (1983)
6.54 T. Schober: In *Hydrogen in Disordered and Amorphous Solids*, ed. by G. Bam-

bakidis, R.C. Bowman, Jr., Proc. NATO ASI Ser.B, Vol.136 (Plenum, New York 1985) p.377
6.55 R. Lässer: J. Less-Common Metals **131**, 263 (1987)
6.56 R. Lässer, K. Bickmann, H. Trinkaus, H. Wenzl: Phys. Rev. B **34**, 4364 (1986)
6.57 O. Blaschko, G. Ernst, P. Fratzl, G. Krexner, P. Weinzierl: Phys. Rev. B **34**, 4985 (1986)
6.58 O. Blaschko, G. Ernst, P. Fratzl, G. Krexner, P. Weinzierl: J. Nucl. Mater. **141-143**, 540 (1986)
6.59 R.O. Simmons, R.W. Balluffi: Phys. Rev. **117**, 52 (1960)
6.60 T. Schober, C. Dieker, H. Trinkaus: to be published
6.61 Zh. Qi, J. Völkl, R. Lässer, H. Wenzl: J. Phys. F **13**, 2053 (1983)
6.62 R. Lässer, to be published in J. Nucl. Mater.
6.63 P. Jung, R. Lässer: Phys. Rev. B **37**, 2844 (1988)
6.64 J.N. Daou, P. Vajda, A. Lucasson, J.P. Burger: Phil. Mag. A **53**, 611 (1986)
6.65 G. Kögel, R. Lässer, W. Triftshäuser, to be published
6.66 P. Hautojärvi (ed): *Positrons in Solids*, Topics Curr. Phys., Vol.12 (Springer, Berlin, Heidelberg 1979)
6.67 T. Schober, C. Dieker, H. Trinkaus: to be published

Subject Index

MIX
Papier aus verantwortungsvollen Quellen
Paper from responsible sources
FSC® C105338

If you have any concerns about our products,
you can contact us on
ProductSafety@springernature.com

In case Publisher is established outside the EU,
the EU authorized representative is:
Springer Nature Customer Service Center GmbH
Europaplatz 3, 69115 Heidelberg, Germany

Printed by Libri Plureos GmbH
in Hamburg, Germany